Advances in Intelligent Systems and Computing

Volume 878

Series editor

Janusz Kacprzyk, Polish Academy of Sciences, Warsaw, Poland
e-mail: kacprzyk@ibspan.waw.pl

The series "Advances in Intelligent Systems and Computing" contains publications on theory, applications, and design methods of Intelligent Systems and Intelligent Computing. Virtually all disciplines such as engineering, natural sciences, computer and information science, ICT, economics, business, e-commerce, environment, healthcare, life science are covered. The list of topics spans all the areas of modern intelligent systems and computing such as: computational intelligence, soft computing including neural networks, fuzzy systems, evolutionary computing and the fusion of these paradigms, social intelligence, ambient intelligence, computational neuroscience, artificial life, virtual worlds and society, cognitive science and systems, Perception and Vision, DNA and immune based systems, self-organizing and adaptive systems, e-Learning and teaching, human-centered and human-centric computing, recommender systems, intelligent control, robotics and mechatronics including human-machine teaming, knowledge-based paradigms, learning paradigms, machine ethics, intelligent data analysis, knowledge management, intelligent agents, intelligent decision making and support, intelligent network security, trust management, interactive entertainment, Web intelligence and multimedia.

The publications within "Advances in Intelligent Systems and Computing" are primarily proceedings of important conferences, symposia and congresses. They cover significant recent developments in the field, both of a foundational and applicable character. An important characteristic feature of the series is the short publication time and world-wide distribution. This permits a rapid and broad dissemination of research results.

Advisory Board

More information about this series at http://www.springer.com/series/11156

Michel Alexandre Cardin
Daniel Hastings · Peter Jackson
Daniel Krob · Pao Chuen Lui
Gerhard Schmitt

Editors

Complex Systems Design & Management Asia

Smart Transportation: Proceedings
of the Third Asia-Pacific Conference
on Complex Systems Design & Management,
CSD&M Asia 2018

 Springer

Editors
Michel Alexandre Cardin
CESAMES Singapore
Singapore, Singapore

Daniel Hastings
Singapore MIT Alliance for Research
and Technology
Singapore, Singapore

Peter Jackson
Singapore University of Technology
and Design (SUTD)
Singapore, Singapore

Daniel Krob
CESAMES
Paris, France

Pao Chuen Lui
National Research Foundation
Singapore, Singapore

Gerhard Schmitt
Singapore-ETH Centre
Singapore, Singapore

ISSN 2194-5357   ISSN 2194-5365   (electronic)
Advances in Intelligent Systems and Computing
ISBN 978-3-030-02885-5   ISBN 978-3-030-02886-2   (eBook)
https://doi.org/10.1007/978-3-030-02886-2

Library of Congress Control Number: 2018959428

This Springer imprint is published by the registered company Springer Nature Switzerland AG
The registered company address is: Gewerbestrasse 11, 6330 Cham, Switzerland

Preface

Introduction

This volume contains the proceedings of the Third International Asia-Pacific Conference on "Complex Systems Design & Management" (CSD&M Asia 2018; see the conference Web site: http://www.2018.csdm-asia.net for more details).

The CSD&M Asia 2018 conference was jointly organized on December 6–7, 2018, at the National University of Singapore (NUS) by the two following founding partners:

1. CESAM Community managed by the Center of Excellence on Systems Architecture, Management, Economy & Strategy (CESAMES),
2. The National University of Singapore (NUS).

The conference also benefited from the permanent support of the two following departments to organize the conference: SMART, the Singapore-MIT Alliance for Research and Technology and SEC, the Singapore-ETH Centre.

We are grateful to many other institutions—academic and professional—that helped us a lot through their involvement during the one-year preparation of CSD&M Asia 2018: DSTA, IRT SystemX, ST Engineering, SUTD, Thales Solutions Asia, and also the International Council on Systems Engineering (INCOSE) who strongly supported our communication efforts.

Many thanks, therefore, to all of them.

Why a CSD&M Asia Conference?

Mastering complex systems requires an integrated understanding of industrial practices as well as sophisticated theoretical techniques and tools. This explains the creation of an annual *go-between* forum in the Asia-Pacific area dedicated to both academic researchers & industrial actors working on complex industrial systems architecture, modeling & engineering. Facilitating their *meeting* was actually for us a *sine qua non* condition in order to nurture and develop in the Asia-Pacific region the new emerging science of systems.

The purpose of the conference on "Complex Systems Design & Management Asia" (CSD&M Asia) is exactly to be such a forum. Its aim, in time, is to become *the* Asia-Pacific academic–government–industrial conference of reference in the field of complex industrial systems architecture and engineering. This is a quite ambitious objective, that we think possible to achieve, based on the success of the CSD&M conference in Paris since 2010 and in Singapore in 2014 and 2016.

Our Core Academic—Industrial Dimension

To make the CSD&M Asia conference a convergence point for academic, government, and industrial communities interested in complex industrial systems, we based our organization on a principle of *parity* between academics, governmental agents, and industrialists (see the conference organization sections). This principle was first implemented as follows:

- Program Committee consisted of a mix between academics and governmental agents/industrialists,
- Invited Speakers came from numerous professional environments.

The set of activities of the conference followed the same principle. They indeed consist of a mixture of research seminars and experience sharing, academic articles, governmental and industrial presentations, etc. The conference topics cover the most recent trends in the emerging field of complex systems sciences and practices from an industrial, governmental, and academic perspective, including the main industrial and public domains (aeronautics & aerospace, defense & security, electronics & robotics, energy & environment, health & welfare services, media & communications, software & e-services, transport, technology & policy), scientific and technical topics (systems fundamentals, systems architecture & engineering, systems metrics & quality, systems modeling tools), and system types (transportation systems, embedded systems, software & information systems, systems of systems, artificial ecosystems).

The Third CSD&M Asia 2018 Edition

The CSD&M Asia 2018 edition received 20 submitted papers, out of which the Program Committee selected 9 regular papers to be published in the proceedings edited by Springer-Verlag. This corresponds to a 45% acceptance ratio which enables to guarantee the quality of the presentations.

Each submission was assigned to at least two Program Committee members, who carefully reviewed the papers. These reviews were managed using the EasyChair conference management system. Our sincere thanks go to Professor Peter Jackson from Singapore University of Technology and Design (SUTD), whose help was precious during this evaluation step.

We also invited eight outstanding speakers from various industrial, governmental, and scientific background, who gave a series of invited talks covering all the spectrum of the conference during the two days of CSD&M Asia 2018. The conference was organized around a common topic: "Smart Transportation". Each day proposed mix invited keynote speakers' presentations and contributed talks (papers accepted by the Program Committee following the call for papers).

August 2018

Michel Alexandre Cardin
Daniel Hastings
Peter Jackson
Daniel Krob
Pao Chuen Lui
Gerhard Schmitt

Acknowledgements

Finally, we would like to thank all members of the Program and Organizing Committees for their time, effort, and contributions to make CSD&M Asia 2018 a top-quality conference. Special thanks are addressed to the CESAM Community team (see http://cesam.community/en) and the National University of Singapore who permanently and efficiently managed all the administration, logistics, and communications of the CSD&M Asia 2018 conference.

The organizers of the conference are also grateful to all sponsors and partners without whom CSD&M Asia 2018 would simply not exist.

Organization

Conference Chairs

General Chairs

Daniel Krob	Ecole Polytechnique & CESAMES, France
Pao Chuen Lui	Adviser, Prime Minister's Office, Singapore

Organizing Committee Co-chairs

Michel Alexandre Cardin	CESAMES Singapore
Daniel Hastings	SMART, Singapore
Gerhard Schmitt	SEC, Singapore

Program Committee Chair

Peter Jackson	SUTD, Singapore

Program Committee

The Program Committee consists of 21 members (academic, industrial, and governmental) who are personalities of high international visibility. Their expertise spectrum covers all the conference topics. The members of this committee are in charge of rating the submissions and selecting the best of them for the conference.

Chair

Peter Jackson   SUTD, Singapore

Members

Steffen Blume   SEC, Singapore
Petter Braathen   Phoenix Consulting AS
Aakil Mohammad Caunhye   NUS, Singapore
Wyean Chan   Université de Montréal, Canada
Eng Seng Aaron Chia   NUS, Singapore
Stefano Galelli   SUTD, Singapore
Daniel Hastings   SMART, Singapore
Paulien Herder   TU Delft, the Netherlands
Elizaveta Kuznetsova   NUS, Singapore
Serge Landry   Thales Solutions Asia, Singapore
Karthik Natarajan   SUTD, Singapore
William Nuttall   The Open University, UK
Ramakrishnan Raman   INCOSE Asia-Oceania
Eng Yau Pee   DSTA, Singapore
Yip Yew Seng   INCOSE, Singapore
Afreen Siddiqi   MIT, USA
Seiko Shirasaka   Keio University, Japan
Kristin L. Wood   SUTD, Singapore
Yixin Jiang   NUS, Singapore
Sixiang Zhao   NUS, Singapore

Organizing Committee

The Organizing Committee consists of 11 members (academic, industrial, and
governmental) in charge of the program and the logistical organization of the
conference.

Co-chairs

Michel Alexandre Cardin   CESAMES Singapore
Daniel Hastings   SMART, Singapore
Gerhard Schmitt   SEC, Singapore

Members

Saik Hay Fong	ST Engineering, Singapore
Peter Jackson	SUTD, Singapore
Hervé Jarry	Thales Solutions Asia, Singapore
Daniel Krob	Ecole Polytechnique & CESAMES, France
Serge Landry	Thales Solutions Asia, Singapore
François Xavier Lannuzel	IRT SystemX, Singapore
Pao Chuen Lui	Prime Minister's Office, Singapore
Yang How Tan	DSTA, Singapore

Invited Speakers

Michael Gastner	Assistant Professor for Mathematics, Computer Science and Statistics, Yale-NUS College, Singapore
Marta C. Gonzalez	Associate Professor, UC Berkeley College of Environmental Design, USA
Mun Leong Liew	Chairman, Changi Airport Group, Singapore
Hoon Ping Ngien	President, Land Transport Authority, Singapore
Yoshiaki Ohkami	Executive Advisor for Institute of System Design and Management, Keio University, Japan
Nelson Quek	Head of Engineering Division, Port of Singapore Authority, Singapore
Ravinder Singh	President, Electronics, ST Engineering, Singapore
Peng Yam Tan	Chief Executive, DSTA, Singapore

Contents

A Net-Based Formal Framework
for Causal Loop Diagrams

Guillermina Cledou[1] and Shin Nakajima[2(✉)]

[1] HASLab INESCTEC & University of Minho, Braga, Portugal
mgc@inesctec.pt
[2] National Institute of Informatics, Tokyo, Japan
nkjm@nii.ac.jp

Abstract. Causal Loop Diagrams (CLDs) are a modeling tool employed in Business Dynamics. Such a diagram consists of many tightly coupled loops to capture dynamic behavior of systems. Intuitive operational semantics, describing how changes are propagated among the loops, provide a basis for model animation or manual inspection. They are, however, not precise enough to enable automated property checking. This paper proposes and defines a net-based formal framework, showing true concurrency, so that automated analysis is made possible.

1 Introduction

Smart transportation is a strategic area to focus on enabling sustainable societies. Managing such large-scale complex systems is mandatory, which needs their modeling from various viewpoints. Ontology captures common vocabulary with its focus on structural aspects of domain concepts (e.g. [2]). System Dynamics (SD) approach is elucidating dynamic behavior (e.g. [9]). The two approaches are complementary, and dynamics are often complex and tightly coupled implicitly, leading to unknown deficiencies. Removing vulnerabilities is essential to avoid *normal accidents* [8]. For example, safety analysis in Aerospace engineering [5] is one successful application of SD to study how to prevent disastrous accidents.

Modeling dynamic behavior of systems may start with qualitative causal loop diagrams (CLDs) followed by a quantitative Stock-Flow approach [10]. A CLD illustrates causal links between concepts, and brings out *mental images* of various stakeholders involved, describing system structure and a hypothesis on its dynamic behavior. A modeling goal is representing complex systems, and thus a straight forward style of constructing a whole CLD is infeasible. Building a CLD is a part of an iterative modeling process, and involves trial-and-error steps, interleaving of description and analysis, before obtaining a satisfactory CLD. Because a CLD is basically representing dynamic behavior, manual inspection is cumbersome as a CLD becomes large. Automated analysis tools can bring great benefits at reduced cost, which is recognized well in modeling of software at an abstract level [3]. Unfortunately, definitions of CLD [10] are not precise

© Springer Nature Switzerland AG 2019
M. A. Cardin et al. (Eds.): CSD&M 2018, AISC 878, pp. 1–12, 2019.
https://doi.org/10.1007/978-3-030-02886-2_1

enough to enable systematic automated analysis such as model checking [1,7]. This hinders from adapting a tool-supported iterative process of building CLDs.

The present paper proposes a new net-based formal framework, *causal loop net (CLN)*, as a basis of formal representation and analysis of CLD. CLN is inspired by Petri-nets [6], especially CP-nets [4], because of their *true concurrency* characteristics, but introduces notions of abstractions to take into account the qualitative nature of CLD. Main contributions of this paper are introducing a new net-based formal framework CLN, and demonstrating the ideas with a proof-of-concept (PoC) tool written in Scala.

The reminder of the paper is organized as follow. Sections 2 and 3 recall CLDs and Petri-nets. Section 4 summarizes issues and our approach to solving them. Section 5 proposes a formal definition of CLN, and then Sect. 6 discusses ways to specify properties to check. Section 7 demonstrates example cases. Section 8 concludes the paper.

2 Causal Loop Diagrams

2.1 Diagram Representation

Causal loop diagrams (CLDs) are a modeling tool to qualitatively represent feedback control systems of non-linear dynamics (Chap. 5 in [10]). A CLD illustrates variables connected via causal links to form loops. Each causal link refers to a fact that a source variable affects a variable at a destination end. Causal links are annotated with either a positive (+) or negative (−) polarity.

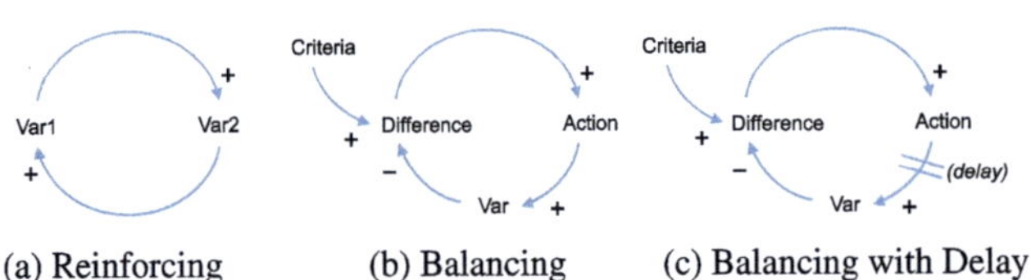

(a) Reinforcing   (b) Balancing   (c) Balancing with Delay

Fig. 1. Three basic components

CLD offers three basic components. Figure 1(a) is a loop in which two connected variables reinforce each other; if `Var1` increases, then `Var2` increases, and similarly for cases of decreasing. Figure 1(b) illustrates a situation where two variables `Difference` and `Var` are balancing; if `Var` increases, then `Difference` decreases because of the negative polarity on the link, and similarly for cases of decreasing `Var`. A balancing loop represents a negative feedback. Figure 1(c) is an example component to have a delay annotation on a causal link between `Action` and `Var`. Changes are not propagated immediately, but deferred, which gives systems inertia, posing much effects on dynamics. These basic components are combined to form large loops to represent complex system dynamics. Figure 2 is an example to have feedback control on a variable `WP` via two loops.

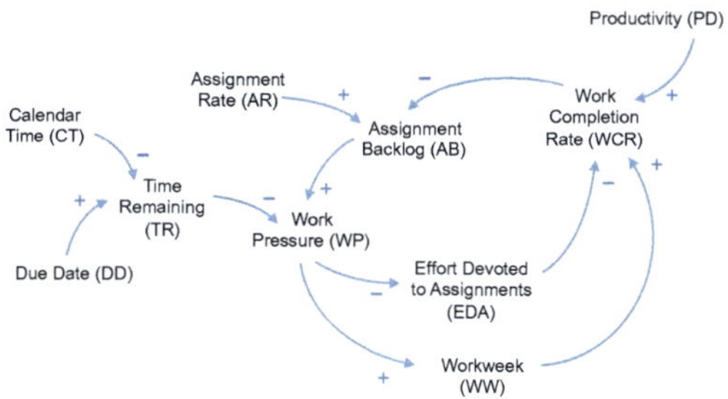

Fig. 2. An example composed loop (Figs. 5–21 in [10])

2.2 Informal Semantics

Structural aspects of a CLD are basically captured by a four-tuple $\langle V, I, \ell, \tau \rangle$. V is a finite set of variable names, and I is an initialization relation. ℓ and τ, representing causal links, are transfer relations of $V \times Polarity \times V$ where $Polarity \stackrel{def}{=} \{+, -\}$. ℓ are simple causal links while τ are those with a delay annotation. They are disjoint; $\ell \cap \tau = \emptyset$. Elements of $Polarity$ are $sign(\partial v_d / \partial v_s)$ when a causal link directs toward a destination v_d from a source v_s ($v_d, v_s \in V$).

Dynamic behavior of a CLD is a set of possible sequences of *fired* transfer relations. Let S_t be a set of *enabled* transfer relations at a time point t in a global time-line. Enabled transfer relations can be fired to propagate changes in a source variable to a destination. Get such a relation r from S_t. If $r \in \ell$, r is fired immediately to propagate changes according to the above mentioned rules on the polarity. If $r \in \tau$, r is rewritten to an ℓ form, and is added to S_{t+d} for a certain d ($d > 0$). It effectively makes r to be fired after passing d ticks; namely r is *delayed*.

CLD defines a notion of being *fired* in an intuitive manner. Tentatively we may have a naive view of starting with initialized terminal source variables and tracing changes in all the variables along causal links. As a number of causal links together with variables becomes large, inspecting these changes manually is cumbersome. Although, for example, the CLD of Fig. 3 in [9] is not large, how the variables change their values is not intuitively clear. Some automated analysis methods are desirable.

3 Net-Based Formal Frameworks

Petri-nets [6] is a net-based formal framework for modeling concurrent systems. A Petri-net is a weighted directed bipartite graph, $\langle P, T, F, W, M_0 \rangle$. P is a finite set of *places*, and T is a finite set of *transitions*, both of which constitute nodes of a graph. F is a flow, a subset of $(P \times T) \cup (T \times P)$, describing edges between nodes. W is a weight function[1] $F \longrightarrow \mathcal{N}$. M_0 is an initial marking discussed

---

[1] $\mathcal{N}$ is a set of Natural Number and $\mathcal{N}_0$ includes 0.

below. A place can hold more than one *token* and thus is a multi-set or a bag of tokens. Tokens are indistinguishable with each other. Figure 3 illustrates a simple example Petri-net, actually a Place/Transition net (PT-net), in which • refers to a token.

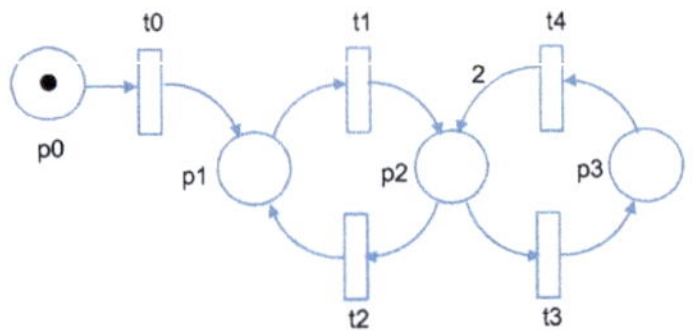

Fig. 3. Petri net

Dynamic behavior of a PT-net is defined in terms of *markings*. Let $M(p)$ represent a multi-set of tokens at a place p. Informally, a transition t is fired when its input places contain enough number of tokens to enable the transition. Furthermore, more than one transition can be fired at the same time. It is referred to as *true concurrency*, which can be compared with interleaving semantics adapted in most of concurrent computation models.

Figure 3 shows a simple example PT-net[2] shows that an initial marking $M_0 = \{p0 \mapsto \{\!| \bullet |\!\}\}$. Transitions of markings, starting from M_0, constitute reachability graphs, which are basis of analyzing dynamical aspects of PT-net. Reachability graphs are graphs of markings, and thus are called *marking graphs* in this paper.

In the example PT-net in Fig. 3, starting with M_0, the transitions $t0$ and $t1$ can fire consecutively. A resulting new marking ($M_2 = \{p2 \mapsto \{\!| \bullet |\!\}\}$) may enable two transitions $t2$ and $t3$. The two are in *conflict*; namely, only one of them can fire even if both are enabled. Assume that $t3$ is fired, which leads to $M_4 = \{p2 \mapsto \{\!| \bullet,\bullet |\!\}\}$ after firing $t4$ whose output edge has a weight annotation of 2. As this M_4 has two tokens, two enabled transitions $t2$ and $t3$ can fire at the same time, showing true concurrency. This results in $M_5 = \{p1 \mapsto \{\!| \bullet |\!\}, p3 \mapsto \{\!| \bullet |\!\}\}$. A transition sequence may continue to a situation where $M_9 = \{p2 \mapsto \{\!| \bullet,\bullet,\bullet |\!\}\}$. The number of tokens is increasing. This PT-net is unbounded, and its reachability graphs are infinite.

Apart from the basic PT-net, various high-level Petri-nets have been proposed so far, including Coloured Petri Nets (or CP-nets) [4]. CP-nets can work on colored tokens. Tokens with different colors are distinguishable, and transitions may have guard conditions on token colors; only tokens with a certain color can enable such transitions. Intuitively, a marking in CP-nets M is partitioned into color-indexed markings M^c. Occurrence graphs of CP-nets are defined similar to reachability graphs of PT-nets.

Note that analysis methods using either reachability graphs or matrix-equations are studied for PT-nets [6]. Flow of tokens can be represented uniformly with matrix-equations, because tokens are indistinguishable. However,

---

[2] This is an infinite capacity net, in which each place can hold any number of tokens.

analyzing CP-nets is possible only with occurrence graphs [4] because tokens with different colors are distinct.

4   Abstractions Qualitatively

CLD represents system dynamics qualitatively, and introducing qualitative abstractions is one of key issues. We will study abstractions from three viewpoints.

4.1   Qualitative Values

As presented in Sect. 2, variables do not take values, but represent qualitatively an increase or a decrease. We, however, introduce a notion of values that a variable can take, to make defining operational semantics easy.

Firstly, we view that a variable v_X in CLD may stand for δX of an accompanying hypothetical variable X. Secondly, we introduce a set Q to be $\{up,\ down\}$, and an extension $\hat{Q} \stackrel{def}{=} Q \cup \{none\}$. The values up and $down$ refer to an increase and a decrease respectively, and $none$ stands for being unknown or no change. A function rev, of $\hat{Q} \longrightarrow \hat{Q}$, is defined such that $rev(up) = down, rev(down) = up$, and $rev(none) = none$.

For a variable $v \in V$, $[\![\ \_\ ]\!]$ is a dereference function, of $V \longrightarrow \hat{Q}$. An element in an initialization I specifies an initial value of a variable v_i ($[\![v_i]\!] \in Q$). Variables not appeared in I take $none$ as their initial values. Transfer relations specify how a value in a source v_s is transferred to a destination v_d, and an annotated polarity affects the transferred values. With $[\![v_s]\!] \in Q$, the equations below illustrates how the polarity acts in the transfer.

$$[\![\ v_d\ ]\!] = [\![\ v_s\ ]\!] \quad \textbf{if } \ell(v_s, +, v_d), \qquad [\![\ v_d\ ]\!] = rev([\![\ v_s\ ]\!]) \quad \textbf{if } \ell(v_s, -, v_d)$$

Each element of $\hat{Q}$ may be encoded as a distinct color if we use CP-nets as a formal framework to encode CLDs. CLD may be encoded as a kind of CP-nets.

4.2   True Concurrency and AMAN Strategy

As presented in Sect. 3, Petri-nets, either CP-nets or PT-nets, can represent true concurrency in a faithful manner. However, the example scenario of Fig. 3 illustrates a situation that places may have an infinite number of tokens, which is clearly not desirable from a view point of automated analysis methods. This infiniteness comes from characteristics of tokens in Petri-nets.

Tokens in Petri-nets can have more than one role. Firstly, a token is representing a computation thread, and thus multiple tokens can refer to multi-threaded computations. Secondly, a token is a single computing resource, and moving a token from a place to another illustrates a situation where an existing resource is consumed at a source place and a new one is produced at a destination.

CLD is abstract, and does not have any notion of computing resources, but its true concurrency aspects are essential. This observation leads to an idea that we duplicate tokens as many as needed, which might be called an *AMAN strategy* in this paper, so as to satisfy desirable degrees of concurrency. While a place in Petri-nets is a multi-set of tokens, we use a set of tokens instead. In addition, we adapt an AMAN strategy to duplicate tokens in a place (p) to fire all transitions connected outwards from p. Note that with an AMAN strategy, no *conflict* occurs (cf. Fig. 3). We abstract multi-sets of tokens to be sets of tokens, which ignores the computing resource aspects of tokens. CLN is different from CP-nets.

4.3 Non-deterministic Delay

Delay is quantitative in nature (Chap. 11 in [10]). A delayed causal link in qualitative CLD is an abstraction of *stock* of quantitative Stock-Flow models. Introducing a delay distinguishes an event of writing values to a stock from another event of reading values. Because these two events are distinct, not occurred at the same time, CLD adapts an abstraction such that transfer values is *delayed*, but a delay is not accompanied with any quantitative amount of time.

Delay makes much effects on a possible transition sequences, and thus on dynamic behavior of CLD [10]. Intuitively, when a particular causal link t_d is annotated with delay (τ), the link is not fired immediately even if true concurrency semantics allow firing the transition transfer of t_d. Now, compare two cases where a particular transition t is in τ (delay) and it is in ℓ (simple). Possible transition sequences of the second simple case is different from the first. Firing t, in the first delay case, appears later in transition sequences than the second. Furthermore, such transition sequences involving delayed t_d might be different for different quantitative amount of delay times. As CLD does not refer to quantitative values, we are not able to specify particular situations selectively.

We introduce a notion of non-deterministic delay time as an abstraction. Given an arbitrary upper bound $d^{upper} \in \mathcal{N}$, we choose a delay time d ($d < d^{upper}$) in a non-deterministic manner when constructing marking graphs. Marking graphs obtained in this way are different for different chosen values of d. Covering all possible cases is impossible, and thus their dynamic behavior is under approximation, namely searching in a part of all possible sequences only.

5 Causal Loop Nets

Causal loop net (CLN) is a directed bipartite graph that we propose in this paper as a formal framework of CLD. CLN is structurally isomorphic to CLD (Fig. 4), in which places of CLN correspond to variables of CLD. We may consider CLN as an internal representation of CLD; modelers need not care about CLN.

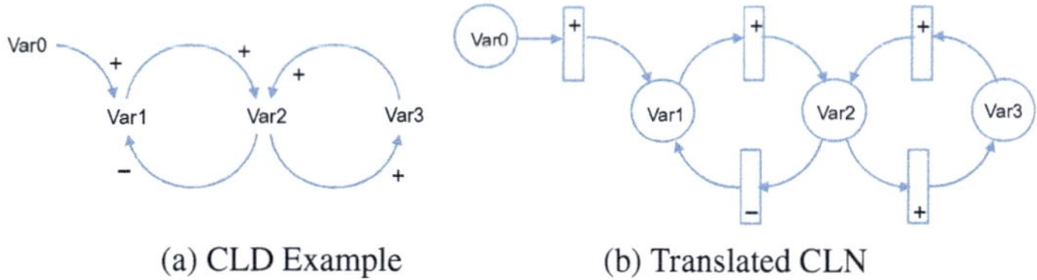

(a) CLD Example   (b) Translated CLN

Fig. 4. CDL and CLN

5.1 Formal Definitions

A CLN is a six-tuple D over $Token$, $D = \langle P, T, F, \tau, \nu, M_0 \rangle$. P and T are finite sets of places and transitions respectively, and $F \subset (P \times T) \cup (T \times P)$. τ is a polarity annotation, $T \longrightarrow \{+, -\}$. ν is a delay annotation, $T \longrightarrow \mathcal{N}_0$. M_0 stands for an initial marking, which will be explained below. For a transition t, a set of input places ${}^\bullet t$ and a set of output places $t^\bullet$ are defined.

$${}^\bullet t = \{\, p \mid (p, t) \in F \,\}, \qquad\qquad t^\bullet = \{\, p \mid (t, p) \in F \,\}$$

Tokens in CLD are either basic tokens or delay tokens; $BasicToken = \{\uparrow, \downarrow\}$ and $DelayToken = \{\uparrow_d, \downarrow_d\}$ for $d \in \mathcal{N}$, where $\uparrow$ is up and $\downarrow$ is $down$. We introduce a whole set of tokens $Token$ to be $BasicToken \cup DelayToken \cup \{\_\}$, where $\_$ is $none$. d in $\uparrow_d$ or $\downarrow_d$ refers to an amount of delay if $d > 0$. For a technical reason, we assume the relationships that $x_0 = x$ for a token x; namely, $\uparrow_0 = \uparrow$ and $\downarrow_0 = \downarrow$. Intuitively, a delay token x_d decreases its amount one as ticks proceed, reaching x_0 to contribute enabling of transitions because $x_0 = x$. Now, $Token$ can be simplified to be $\{\uparrow_e, \downarrow_e, \_\}$ for $e \in \mathcal{N}_0$. A marking M is $P \rightarrowtail 2^{Token}$.

Given a marking M, a transition t is enabled if its all input places have tokens other than $\_$ ($none$). Thus, a set of enabled transition is

$$\mathcal{T}(M) = \{\, t \in T \mid \textstyle\bigwedge_{p \in {}^\bullet t} M(p) \cap \{\uparrow_e, \downarrow_e\} \neq \emptyset \,\}$$

$\mathcal{T}(M)$ may have more than one transition, which ensures that CLN is a formal framework allowing *true concurrency*.

5.2 Dynamic Behavior

Dynamic behavior of a CLN is captured by an associated marking graph, which changes in markings $(M \xrightarrow{\mathcal{T}(M)} M')$ define. Given a set of enabled transitions $\mathcal{T}(M)$, $M'(p)$ for all places p in P is defined below, $\bigoplus$ denoting *overriding*.

$$M'(p) \;=\; M(p) \oplus \left(\textstyle\bigcup_{t \in \mathcal{T}(M)} {}^\bullet\!\Delta(p, t) \;\cup\; \bigcup_{t \in \mathcal{T}(M)} \Delta^\bullet(t, p)\right)$$

where ${}^\bullet\Delta$ and $\Delta^\bullet$ are of $F \longrightarrow 2^{Token}$. Recall that $F \subset (P \times T) \cup (T \times P)$.
${}^\bullet\Delta(p, t)$ removes all the tokens in $M(p)$ and leaves delay tokens whose residual time is decremented by one.

$$\bullet\Delta(p,t) \;=\; \bigcup_{x\in M(p)} Decr(x)$$

$Decr(x)$ makes use of the relation that $\uparrow = \uparrow_0$ and $\downarrow = \downarrow_0$.

$$Decr(x_e): \; Token \longrightarrow 2^{Token} = \begin{cases} \{\, x_{e-1}\,\} & \textbf{if} \quad e > 0 \\ \emptyset & \textbf{if} \quad e = 0 \end{cases}$$

$\Delta^\bullet(t,p)$ adds tokens calculated according to the polarity $\tau(t)$, where $M(\bullet t)$ is an abbreviation of $\bigcup_{p'\in\bullet t} M(p')$.

$$\Delta^\bullet(t,p) \;=\; \bigcup_{x\in M(\bullet t)} \mathrm{Xfer}(x)$$

$\mathrm{Xfer}(x)$ assumes, for simplicity, that $\nu(t) = 0$ if a causal link is not marked *delay*.

$$\mathrm{Xfer}(x): \; Token \longrightarrow 2^{Token} = \begin{cases} \{\, x_{\nu(t)}\,\} & \textbf{if} \;\; \tau(t) = + \\ \{\, rev(x_{\nu(t)})\,\} & \textbf{if} \;\; \tau(t) = - \end{cases}$$

The transition relationship $M \xrightarrow{\mathscr{T}(M)} M'$ may lead to a situation where, for a certain p, $\{p \mapsto \uparrow, \; p \mapsto \downarrow\} \in M'$ occurs. Such a marking is *inconsistent* in that it assigns two tokens $\uparrow$ and $\downarrow$ to a place p at the same time. P_{NG} is a subset of P to contain those inconsistent places; $P_{NG} \overset{def}{=} \{\, p \mid \{\uparrow,\downarrow\} \subseteq M(p)\,\}$.

The inconsistency demonstrates a situation where both cases, p with $\uparrow$ and p with $\downarrow$ are possible for $\forall p \in P_{NG}$. Therefore, two cases are to be explored non-deterministically[3]. An auxiliary function Nr constructs consistent markings by dividing an inconsistent marking into two cases.

$$Nr(M) \;=\; \begin{cases} Nr(M[p \mapsto D(p,\uparrow)]) \cup Nr(M[p \mapsto D(p,\downarrow)]) & \textbf{if} \;\; p \in P_{NG} \\ \{\, M\,\} & \textbf{if} \;\; P_{NG} = \emptyset \end{cases}$$

where $D(p,x) = (M(p) \cap \{\uparrow_d, \; \downarrow_d\}) \cup \{x\}(d > 0)$. Now that $Post(M)$ is a set of consistent markings reached directly from M when inconsistent markings are considered.

$$Post(M) \;=\; \{\, M' \mid M \xrightarrow{\mathscr{T}(M)} M'' \;\wedge\; M' \in Nr(M'')\}$$

Given a CLD D, a marking graph $\mathscr{G}^D$ of a CLN is a tuple $\langle Node, n_0, Edge\rangle$, where n_0 is an initial node representing the initial marking M_0, $Edge$ is a set of transition relations $M \Longrightarrow M'$ for $M' \in Post(M)$, and $Node$ consists of all markings reached from M_0 via $Edge$. $\mathscr{G}^D$ is finite in particular.

6   Formal Analysis

Once a marking graph is obtained for a given CLN or CLD equivalently, we can conduct formal analyses to study the CLN from various ways.

---

[3] This makes it difficult to define matrix-equations for CLN.

6.1 Verification Problems

As explained before, a place of CLN represents a variable in CLD. Since a node n in $\mathcal{G}^D$ is a marking, $[\![n.v]\!]$, a value of variable v at node n, is $M(v)$ and $[\![n.v]\!] \in \hat{Q}$. Delay tokens, $\uparrow_d$ or $\downarrow_d$, are identified with _ (*none*) as values, because they are transient and do not any definite values.

Let $Trace^D$ be a set of all possible transition sequences σ_i generated by $\mathcal{G}^D$ starting from n_0; $Trace^D = \{\ \sigma_i\ \}$. $\sigma_i(r)$ is an r-th node in σ_i, and $\sigma_i(r..s)$ is a transition sequence starting with $\sigma_i(r)$ and ending with $\sigma_i(s)$. In particular, $\sigma_i(0)$ is n_0 for any sequence.

A verification problem is to check whether at least one transition sequence exists in $Trace^D$ to satisfy a given property Φ taking into account simulation relations $\sim$.

$$\exists \sigma_i \in Trace^D : \ \Phi(\sigma_i) \quad \mathbf{mod} \ \sim$$

The simulation relation $\sim$ is either $\sim_1$ or $\sim_2$ below.

$$\sim_1 : \hat{Q}^N \longleftrightarrow Q \ \text{ is } \ q(q|none)^{N-1} \sim_1 q \text{ for } q \in Q,$$
$$\text{and} \quad \sim_2 : \hat{Q}^N \longleftrightarrow \hat{Q} \ \text{ is } \ q^N \sim_2 q \qquad\qquad \text{for } q \in \hat{Q}.$$

Adapting simulation relations makes it easy to write properties to be checked. Transition sequences σ_i are of $seq(\hat{Q})$. Because $q \in \hat{Q}$ does not refer to a quantitative, concrete value, identifying N consecutive q (q^N) with just a q does not make much effects on qualitative characteristics of sequences. Namely, q can be a summary of q^N. Simulation relation of the first type ($\sim_1$) provides a way to use Q as a basis for this summary, while the second ($\sim_2$) is using $\hat{Q}$ instead. We will see how two simulation relations are effective in Sect. 7.

6.2 Some Query Patterns

We introduce a few typical properties to be checked, all of which take a form of queries on $Trace^D$. A finite mapping f of $[0, k-1] \longrightarrow Q$ for a $k \in \mathcal{N}$ is called *a summary function* when f describes desired sequences of a CLD variable. A predicate $\varphi_0(i, r, v, f)$ becomes *true* if there is at least one transition sequence σ_i in $Trace^D$ whose subsequence starting with r and ending with s satisfies the condition that a sequence of values generated by a variable v in D is simulated by a given summary function f.

$$\varphi_0(i, r, v, f) = \bigwedge\nolimits_{j=0}^{k-1} \exists s_{j+1}.\ (\sigma_i(s_j \ .. \ s_{j+1}).v \sim f(j))$$

where $k = |f|$ (length of f), $r = s_0$ and $s = s_k$.

Some properties employ $\varphi_0(i, r, v, f)$ to define themselves. Firstly, $\varphi_1(v, f)$ returns *true* if there is a transition sequence σ_i to have a subsequence of a certain length in which the specified variable v is simulated by a given summary function f.

$$\varphi_1(v, f) \ = \ \exists i, r : \ \varphi_0(i, r, v, f)$$

Secondly, $\Phi_3(\varphi_1(v^m, f), v^\ell, g^\ell)$ is *true* if a summary function of a variable v^ℓ ($m \neq \ell$) is simulated by g^ℓ in a transition sequence that $\varphi_1(v^m, f)$ satisfies;

$$g^\ell(X) \;=\; \exists s. \; \sigma_i(r \mathbin{..} s).v^\ell.$$

Intuitively, Φ_3 extracts a sequence of v^ℓ values along a transition sequence that v^m satisfies f. Therefore, we can compare how two variables, v^m and v^ℓ, change their values in an obtained interval of r and s.

7 Examples

Figure 5 shows two marking graph examples generated with a Scala-based PoC tool using Graphviz[4] for preparing graphical images. Figure 5(a) is the one for the CLD in Fig. 4. Figure 5(b) is simple because of no inconsistent marking ($P_{NG} = \emptyset$). In these graphs, 1, 0 and −1 stands for ↑, _ and ↓ respectively.

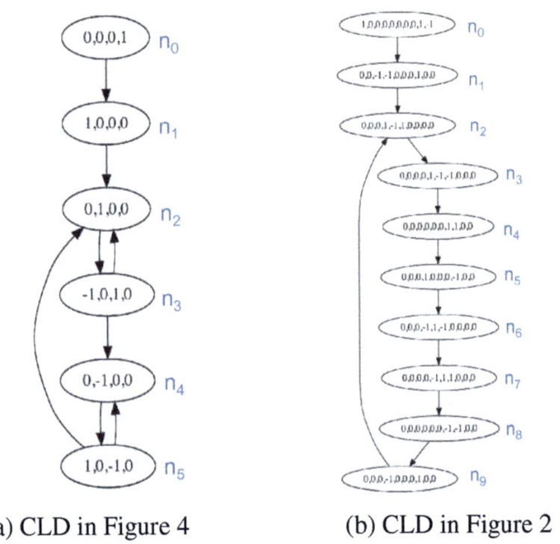

(a) CLD in Figure 4   (b) CLD in Figure 2

Fig. 5. Marking graphs

Each node in the graph of Fig. 5(a) refers to variable values in a form of $\langle$Var1, Var2, Var3, Var0$\rangle$. The graph is the one starting with an initial marking of $M_0 = \{$Var0 $\mapsto$ ↑$\}$. The prototype PoC tool accepts a CLD in a textual form. The CLD in Fig. 4 is, for example, entered as below.

```
var netF4 = newclpn ++ (
   4 >+ 1,
   1 >+ 2,
   2 >- 1,
   2 >+ 3,
   3 >+ 2
) initMark((4, Set(1))
```

---

[4] http://viz-js.com.

The numbers are representing CLD variables. 4 stands for `Var0` whose initial value is 1 (namely $\uparrow$) as specified by `initMark((4, Set(1))`.

We check whether $\mathscr{G}^D$ satisfies $\varphi_1(\texttt{Var2}, f^{\mathrm{Var2}})$ modulo $\sim_1$ where $f^{\mathrm{Var2}} = (\uparrow \uparrow)$. For this, $\varphi_1(\texttt{Var2}, f^{\mathrm{Var2}})$ is *false* because the graph does not have any sequence with consecutive $\uparrow$s. Alternatively, if we choose f^{Var2} to be $(\downarrow \uparrow)$, we actually find transition sequences $\sigma_i(5 \,..)$, some of which are shown here. `Var3` changes its value along each transition sequence.

$$
\begin{aligned}
\sigma_1(5 \,..\, 8) \;&= n_4 \; n_5 \; n_2 \; n_3 & (\sigma_1(5..8)).v^{\mathrm{Var3}} &= {\_} \downarrow {\_} \uparrow \\
\sigma_2(5 \,..\, 10) &= n_4 \; n_5 \; n_4 \; n_5 \; n_2 \; n_3 & (\sigma_2(5..10)).v^{\mathrm{Var3}} &= {\_} \downarrow {\_} \downarrow {\_} \uparrow \\
\sigma_3(5 \,..\, 10) &= n_4 \; n_5 \; n_2 \; n_3 \; n_2 \; n_3 & (\sigma_3(5..10)).v^{\mathrm{Var3}} &= {\_} \downarrow {\_} \uparrow {\_} \uparrow \\
\sigma_4(5 \,..\, 12) &= n_4 \; n_5 \; n_4 \; n_5 \; n_2 \; n_3 \; n_2 \; n_3 & (\sigma_4(5..12)).v^{\mathrm{Var3}} &= {\_} \downarrow {\_} \downarrow {\_} \uparrow {\_} \uparrow
\end{aligned}
$$

A summary function g^{Var3} modulo $\sim_1$ for all the above will be $(\downarrow \uparrow)$, which provides little information. On the other hand, if we use $\sim_2$, the summary function g^{Var3} becomes $((\_ \downarrow)^{n1} (\_ \uparrow)^{n2})$ for $n1, n2 \in [1, 2]$. This scenario illustrates that the $\sim_1$ is appropriate when we extract transition sequences by means of f^{Var2}, and that the $\sim_2$ is useful in searching for summary functions of g^{Var3} because the obtained sequences has detailed information. We may choose g^{Var3} as $(\_ \downarrow)^{n1} (\_ \uparrow)^{n2}$ for $\Phi_3(\varphi_1(\texttt{Var2}, (\downarrow \uparrow)), \texttt{Var3}, g^{\mathrm{Var3}})$ to be satisfied modulo $\sim_2$.

8   Concluding Remarks

Petri-nets have a large body of work [6], from introducing high level Petri-nets such as CP-net, to studying subclasses of PT-nets in view of behavioral or structural characteristics. Causal Loop Net (CLN) is unique in that our approach is introducing qualitative abstractions into a Petri-net family of formal frameworks. CLN provides a formal basis of Causal Loop Diagrams (CLD) to enable formal analysis in terms of marking graphs. Query patterns, although not general enough, are useful in tool-assisted inspections as demonstrated with our PoC tool. The proposed CLN enables adapting a tool-supported iterative process of building CLDs.

Our future plan includes developing a robust tool that can work on large CLDs, studying pros and cons of the proposed abstraction method for delay, and adapting logic model checking methods [1] for formal analyses.

Acknowledgements. The first author conducted the reported work at NII under NII-Internship Program 2017-1 call. This work is a result of project SmartEGOV/NORTE-01-0145-FEDER-000037, supported by Norte Portugal Regional Operational Programme (NORTE 2020), under the PORTUGAL 2020 Partnership Agreement, through the European Regional Development Fund (EFDR). Additional support is provided by the PT-FLAD Chair on Smart Cities & Smart Governance. The second author is partially supported by JSPS KAKENHI Grant Number JP17H01726.

References

1. Clarke, E.M., Grumberg, O., Peled, D.A.: Model Checking. The MIT Press, Cambridge (1999)
2. Cledou, G., Barbosa, L.: An ontology for licensing public transport services. In: Proceedings of the 9th ICEGOV, pp. 230–239 (2016)
3. Jackson, D., Wing, J.: Lightweight formal methods. IEEE Comput. **29**(4), 21–22 (1996)
4. Jensen, K.: Coloured Petri Net. Springer, Heidelberg (1996)
5. Leveson, N.G.: Engineering a Safer World: Systems Thinking Applied to Safety. The MIT Press, Cambridge (2011)
6. Murata, T.: Petri nets: properties, analysis and applications. Proc. IEEE **77**(4), 541–580 (1989)
7. Nakajima, S.: Model checking of energy consumption behavior. In: Proceedings of the 1st CSD&M-Asia, pp. 3–14 (2014)
8. Perrow, C.: Normal Accidents: Living with High-Risk Technologies. Princeton University Press, Princeton (1999)
9. Shepherd, S.P.: A review of system dynamics models applied in transportation. Transp. B Transp. Dyn. **2**(2), 83–105 (2014)
10. Sterman, J.D.: Business Dynamics: Systems Thinking and Modeling for a Complex World. Irwin McGraw-Hill, Boston (2000)

Modeling Operations of a Custom Hiring Center Using Agent Based Modeling and Discrete Event Simulation

Yatin Anil Jayawant[✉] and Nikhil Joshi

Asia Technology Innovation Center, John Deere India Pvt. Ltd.,
Tower 14, Magarpatta City, Hadapsar, Pune 411013, India
{jayawantyatin,joshinikhil}@johndeere.com

Abstract. Custom Hiring Centers (CHCs) for agricultural equipment and services are being promoted by the Government of India and various state governments. CHCs are important part of the complex agricultural system. They are beneficial to the community as small and marginal farmers, instead of having to purchase costly machines, can rent the machines as needed. Key factors that affect profitability are the type and number of equipment hosted by the CHC. Each piece of equipment entails a sunk cost as well as maintenance cost. However, since most agricultural operations are time sensitive, the demand is concentrated in time, and having too few pieces of equipment would entail inability to serve the demand and lost revenue. This paper presents a model to simulate the operations of a CHC, under varying scenarios, in order to estimate its profitability. The model uses a combination of Agent Based Modeling (ABM) and Discrete Event Simulation (DES) approaches. Farmers and equipment are modelled as the agents in the system while the CHC operations are modelled using DES approach. The use of the model is also demonstrated with the help of an example. The results showed that even for one combination of equipment, the profitability varies a lot. This significant variation in profitability arises from variation in the percentages of areas under different types of crop, as well as the inherent variation in crop growth rates, readiness of individual fields for particular operations within the seasonal window, as well as propensity of individual farmers to wait for CHC equipment if not immediately available. Results also demonstrate that simulation is useful to model CHC operations in a complex agricultural system and take informed decision about number and types of equipment to buy.

Keywords: Complex agricultural system
Farm machinery Custom Hiring Center (CHC) · Product mix
Agent Based Modeling (ABM) · Discrete Event Simulation (DES) model
Agricultural economics · Break even analysis · Agricultural start-up
Farm mechanization · Farm machinery services
Farm machinery costs · Multimethod modeling

© Springer Nature Switzerland AG 2019
M. A. Cardin et al. (Eds.): CSD&M 2018, AISC 878, pp. 13–24, 2019.
https://doi.org/10.1007/978-3-030-02886-2_2

1 Introduction

A Custom Hiring Center (CHC), in the context of agricultural equipment, refers to a company or an individual that owns a fleet of equipment and provides equipment (or specific services using those equipment) on rent to farmers as needed. The main advantage of CHCs is that they enable small and marginal farmers, who can't afford to buy equipment outright, reap the benefits of mechanization. Hiremath et al. [1] found that productivity and income of small and marginal farmers increased 10–15% after using machines provided by CHC. CHCs also help in effective use of otherwise expensive machines.

While the benefits of CHCs for small and marginal farmers are accepted, many such centers have struggled to keep their operations profitable. Hiremath et al. [1] also reported that economic utilization of machines in most of the CHCs surveyed was very low. They also found that net returns were also very low, averaging only INR 8822/- ($132.33) per year. In another study Srinivasarao et al. [2] provided information on about 100 CHCs situated in climatically affected regions across India. The farm implements owned by these centers varied from four to 31, and revenue generated varied from nil to INR 145,000/- ($2175) per year. The factors considered to affect revenue generation were equipment selection, center management and socio-economic conditions of the farmers. Sidhu and Vatta [3] conducted a study in Punjab, India to evaluate contribution of Cooperative Agro Machinery Service Centers (AMSCs). Of the total farmers surveyed, 73% farmers suggested to increase machines at AMSCs to have timely availability during peak season. Chahal et al. [4] surveyed 100 Cooperative Agro-Service Centers (CASCs) in order to examine their role in institutionalization of custom hiring services in Punjab, India. They found that there was a vast difference between minimum and maximum profits of CASCs, from INR 2,000 ($30) and INR 770,000 ($11550) in 2008–09 to INR 30,000 ($450) and INR 667,570 ($10013) in 2011–12. Though the authors did not specifically mention reasons behind such a difference, they acknowledged the role of farm size, labor availability and custom services, crop selection and cultural practices on the selection of optimum equipment set and number of equipment. The increase in income over the years was owing to the subsidy given by the government.

Equipment selection while setting up a CHC involves deciding the types of equipment, as well as the number of equipment of each type. The type of equipment is generally decided based on the needs of the crops, soils, and agricultural practices of the particular region. However, deciding the number of equipment of each type is a challenging task. Each additional item entails an upfront cost, as well as costs for maintenance and upkeep. However, since most agricultural operations are time sensitive, the demand for particular equipment or services is concentrated in small periods. Having fewer numbers of any type may entail inability to serve the demand and consequently lost revenue. Thus, the optimal number to purchase depends on a combination of factors that vary for each type of equipment.

A number of studies have been completed to understand economics of owning or operating agricultural machinery. Kamboj et al. [5] have studied seven CHCs and calculated Break Even Point in terms of hours per year of various machines. They also noted the type of equipment and minimum investment needed to be profitable. Paman et al. [6] conducted research to analyze cost, profit and break-even of power thresher hiring services in Kampar Regency, Indonesia. They calculated the break-even point in terms of seasonal work (kg) for different custom hiring rates. Schular and Frank [7] provided equipment life tables along with a worksheet to calculate fixed and operating costs for different machines/equipment. Microsoft Excel worksheets for estimating machinery costs have been created and provided by universities [8] and governments. Singh and Mehta [9] presented a decision support system for calculating the total operating cost and break-even units of farm machinery.

There are also examples from other domains where authors have studied entire fleet of equipment for optimizing number suitable for the purpose. Maisenbacher et al. [10] proposed the use of Agent-Based Modeling to determine the minimum number of resources (e-bikes) required to serve the rental needs of a particular area. Eid [11] proposed a technique called General Utility Simulation System (GUSS) to find out optimum number of vehicles in a service-equipment fleet. Cook [12] in his report for North Carolina Department of Transportation, has done extensive literature study on fleet optimization and devised a spreadsheet based tool to assess relative vehicle use and productivity assessment.

In this paper, we present an approach using a combination of Agent Based Modeling and Discrete Event Simulation to simulate the operations of a CHC given a particular fleet of equipment, and information about the context, such as number of potential customers, time window for specific operations, etc. The simulation is used to estimate the cash flow, and consequently the profit, for the CHC over the period of the simulation. The simulation model is developed using AnyLogic, a multimethod simulation software.

The remainder of the paper is organized as follows: Sect. 2 explains the modeling approach, including considerations and assumptions. Section 3 demonstrates the use of the model with the help of an example. Section 4 presents a summary of the work done, along with some insights and directions for further research.

2   Modeling Approach

As discussed in Sect. 1, the primary purpose of the model is to estimate the profitability of a CHC for a given fleet of equipment. Thereupon different options of fleet sizes can be simulated to choose the size that is likely to be profitable.

In case of CHC, profit is complex function of factors like Crop (cropping pattern, time windows of operations, etc.), Geography (fuel cost, population/customer base, landholding, etc.), equipment parameters (number of equipment, their initial price, rental, field capacity, fuel consumption, etc.), admin cost, and other factors like farmers' patience in getting resources, etc. In the model, profit is calculated by subtracting expenses from revenue, over period of simulation. Where, revenue is obtained by multiplying custom hiring rate (rent) of all equipment with their working time. Expenses are categorized in three types viz. CHC admin cost, self-propelled and implement expenses. Self-propelled equipment (e.g. Tractors) expenses include repair & maintenance, taxes, insurance & housing, interest, and fuel cost (Fuel price * Fuel consumption * working time). Implement expenses include repair & maintenance, taxes, insurance & housing and interest. Machine working time depends on the number of customers/farmers availing services, their landholding, their patience in getting the machine, time window of operation and machine's field capacity.

The model comprises of two main entities, viz. the Custom Hiring Center and a population of farmers (or customers) that rents equipment/services from the center. A set or resource pool is maintained for each type of equipment provided by the center. Equipment are modeled as Agents of type "Tractor" and "Implement". Each "Tractor" and "Implement" has associated economic parameters, such as price, maintenance costs, running costs, diesel consumption rate, etc. A "Tractor" also has some operating parameters, such as its field capacity (i.e. the amount of area it can service per unit time). The "Tractor" and "Implement" behavior is simple, as it merely switches between "idle and available" and "working" states. A schedule of operating hours is defined for each set of equipment or resource pool, which constrains the equipment from being issued for use outside of working hours, even if the equipment is in the "idle and available" state.

The actual day to day operation of the CHC is modelled using a discrete event simulation model, similar to models used for manufacturing process simulations. The process model is shown in Fig. 1. The current model is built for a CHC that provides four types of services, although the methodology can be extended to any number of services. When any request for service is received by the CHC, it is placed in one of the four queues (denoted by the "seizeXX" stages in Fig. 1) based on the type of service requested. The requests are served on a first come first served basis. The CHC may require one or more types of equipment to provide the service. Thus, when a request arrives at the head of the queue, it is issued the required equipment if and only if the required pieces of equipment are in the "idle and available" state and current time is within the working hours. Otherwise, the request continues to remain in queue until the conditions are met. The model also allows for a request to exit the queue without being serviced (as explained in farmer behavior in the following paragraphs). Once the pieces of equipment are issued, the request is considered to be in the process of being serviced (this is shown by "delayXX" stages in Fig. 1). The respective pieces of equipment are also placed in the "working" state. The time for which the request remains in this stage depends upon the size of the field where the service is requested and the field capacity of the equipment. Upon completion of this time, the request is advanced to the completed stage and the respective pieces of equipment are returned to the "idle and available" state. As can be inferred, the number of requests that may be simultaneously serviced is constrained by the number of equipment, of the types required for the request, held by the CHC. A rental rate is also defined for each type of service. During the simulation, the rental rate and duration of service is used to calculate revenue for the CHC.

The economic information for each piece of equipment as well as its utilization are used to calculate fixed expenses (such as down-payment on purchase, monthly installments, maintenance, etc.) as well as variable expenses (such as fuel consumption and consumables) for the CHC. Additional expenses for storage and administration are also calculated to continuously estimate the cash flow for the CHC over the entire simulation.

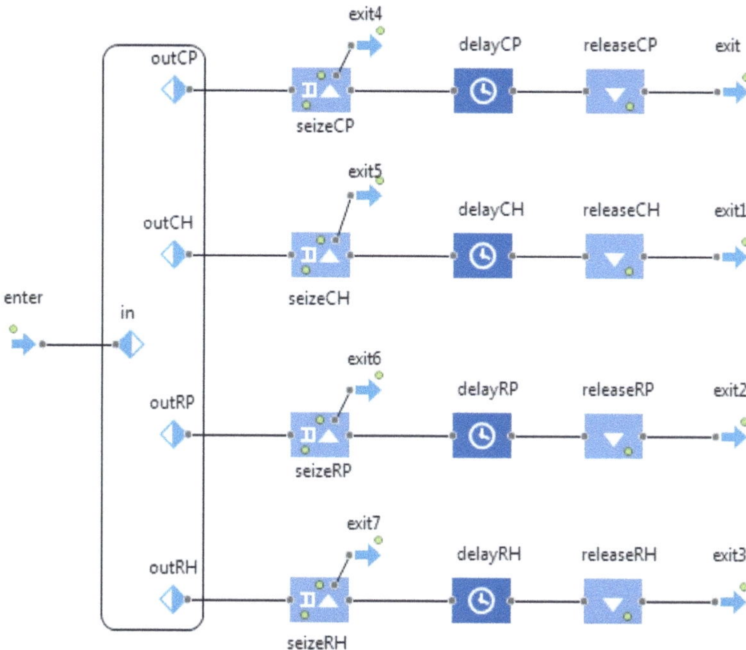

Fig. 1. Process flow chart of Custom Hiring Center operation

The other important component of the model is the population of farmers or customers. Each individual in the population is modeled as an Agent of type "Farmer". The "Farmer" Agent is modeled to simulate the behavior of farmers who will rent equipment from the CHC. Each "Farmer" Agent has a variable to denote the "area of field", and a variable to represent the "current crop". The variable for "area of can be randomly assigned based on a distribution of field sizes in the location which is served by the CHC. This variable remains constant throughout the simulation. The crops grown in the area are stored in a list of enumerated values, and the "current crop" variable can be randomly assigned one of the enumerated values based on the relative prevalence of each crop in the particular season. The "current crop" variable is updated at the end of the corresponding crop cycle.

For each crop, certain agronomic information about the crop cycle, as practiced in the location served by the CHC, is provided. This includes approximate date ranges of operations, such as land preparation or harvesting, for which equipment can be rented from the CHC. At present, only two crops, each having two operations for which equipment can be rented, are defined in the model. However, the model can be easily extended to incorporate additional crops by including them in the list of enumerated values and providing certain agronomic information.

As is the conventional practice in India, it is assumed that there are two main cropping seasons in the year, viz. *Rabi* (winter crop season) and *Kharif* (monsoon crop season). A single crop cycle may span a single season or multiple seasons. However, the next crop is taken only at the start of the next cropping season after the previous crop cycle is completed. Thus, the behavior of the "Farmer" Agent is designed to cycle through different crops as shown in the state chart in Fig. 2.

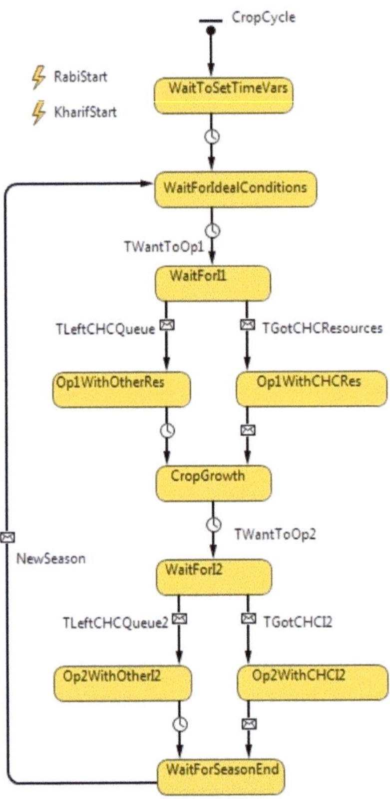

Fig. 2. State chart describing behavior of Farmer Agent

At the start of the simulation, the "Farmer" is assumed to have chosen the "current crop" for growing in the season, and will be waiting for ideal conditions to begin the first operation for which equipment can be rented. The time to transition out of this state is randomly decided for each "Farmer" within the date range provided in the agronomic information provided for the respective crops. Once this transition occurs, the "Farmer" is assumed to request the equipment from the CHC and enters a state of waiting in queue for the equipment to be issued. This state corresponds to the "seizeXX" stage in the process chart of the CHC, as shown in Fig. 1.

Once the "Farmer" is issued the equipment needed, the "Farmer" transitions to the "Operation with CHC Resources" state. The state corresponds to the "delayXX" stage in the process chart of the CHC.

Thus, the "Farmer" remains in this state for a duration that depends on its "area of field" and the "field capacity" of the equipment issued. The behavior defined for the "Farmer" also allows the "Farmer" to exit the queue at the CHC if the "Farmer" has been waiting for a long time. This is denoted by the transition to the state "Operation with Other Resources". The threshold for this transition is randomly decided for each "Farmer" based upon the time window available for the operation and a probability distribution that represents typical patience of a farmer. Upon completion of the operation (either with CHC resources or other resources), the "Farmer" transitions to a state that denotes the period before the next operation for which equipment can be rented. The behavior for this second operation is similar to the behavior for the first operation. The crop cycle is assumed to be completed after the second operation and subsequently, the "Farmer" enters a state of "Wait for season end". Upon the start of the next season, the "current crop" for the "Farmer" is updated based upon crop rotation rules provided for the respective location, and the crop cycle behavior is repeated.

It should be noted that the behavior of each entity in the model is based on predefined rules and that there is no decision making or learning of behavior as the simulation progresses. Events and conditions for state transitions are explicitly calculated and triggered within the model, and hence there is no need for continuous sensing of any environmental parameters. Individual "Farmer" Agents do not communicate amongst each other, and hence factors such as referrals or effect of word of mouth publicity are not incorporated in the model. Communication between "Farmer" Agents and the CHC are simulated by parallel workflows/behaviors that are triggered simultaneously.

3 Implementation Example, Results and Discussion

We demonstrate the use of the model developed with the help of an example. For this example, we assume that the CHC caters to a population of 200 farmers. The "FieldArea" variable is assigned a value such that the areas of all the farmers follow a triangular distribution between 2 and 8 acres with mode at 5 acres. The "current crop" is initialized to one of the two crops with equal probability. It is assumed that each operation requires a specialized implement and a tractor to pull the implement. A specialized implement is shared by both crops for one type of operation. Thus, the CHC has three resource sets of equipment, two sets of special implements and one set of tractors. Table 1 summarizes the equipment combinations used in the example.

Table 1. CHC equipment configurations

Configuration name	No. of tractors	No. of type 1 implements	No. of type 2 implements
Combo 1	2	1	1
Combo 2	4	2	2
Combo 3	6	3	3
Combo 4	7	3	3
Combo 5	8	4	4
Combo 6	10	5	5
Combo 7	11	5	5
Combo 8	12	6	6
Combo 9	13	6	6
Combo 10	14	7	7

For a configuration (6 tractors, 3 implements of type 1 and 3 of type 2), the simulation was run for a period of 5 years. Figure 3 shows the calculated cash flow for the CHC over the period of the simulation for a single simulation run. Trend and seasonality in cash flows can be seen by taking a closer look. This has followed the demand of equipment by farmers for agricultural operations. There is an overall increasing trend in expenses, revenue and net income. Two immediate peaks (days 390 & 480, 750 & 840 and so on) and downward trend till next peak shows a typical agricultural season for given crops. A downward trend within a season also mimics the crop growth period when no operations are carried out and hence no demand for equipment. This chart would be indicative of the profitability of the chosen configuration and the crests and troughs over the years.

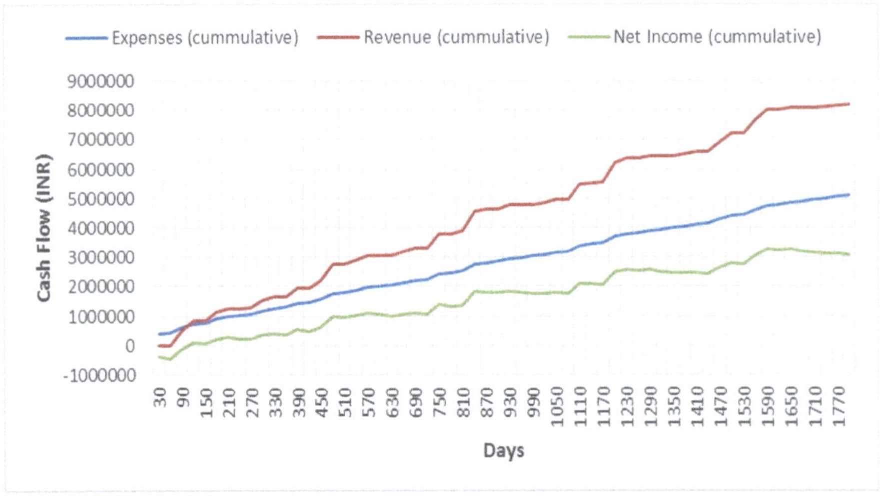

Fig. 3. Calculated cash flow and Break Even Point for example simulation run

The pie chart in Fig. 4 shows the proportion of requests for renting equipment that were served by the CHC, and the requests that left the queue for not receiving the equipment in time. This chart could be used as indicative of demand that can be catered to by having additional numbers of equipment. Such pie charts could be created for each type of equipment to understand which type of equipment has higher unmet demand. It should be noted, however, that Figs. 3 and 4 only indicate the results for the given configuration under one random scenario.

Monte Carlo simulation runs are required to estimate the expected value of profit as well as the range of profitability of a given configuration. For different CHC configurations in the above example, 100 simulation runs were used to calculate the expected profit and variability in the profit. Figure 5 shows the expected profit at the end of 5 years along with 3-sigma limits for 10 different CHC configurations. For the given scenario, profit increased while going from configuration 1 to 2, but then decreased. For configuration 9, average profit is above zero but there are chances of loss too.

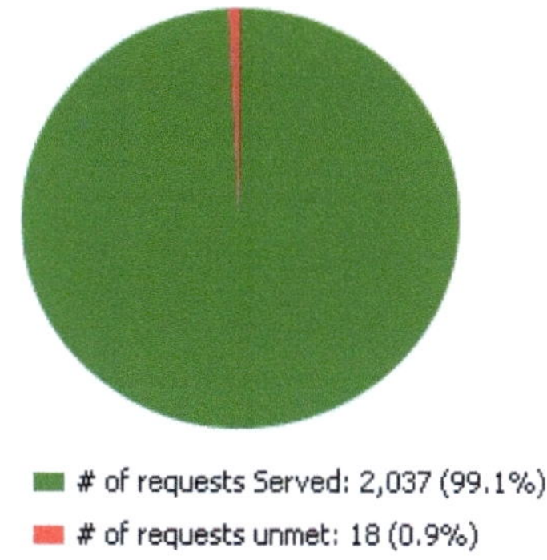

Fig. 4. Proportion of unmet demand for example simulation run

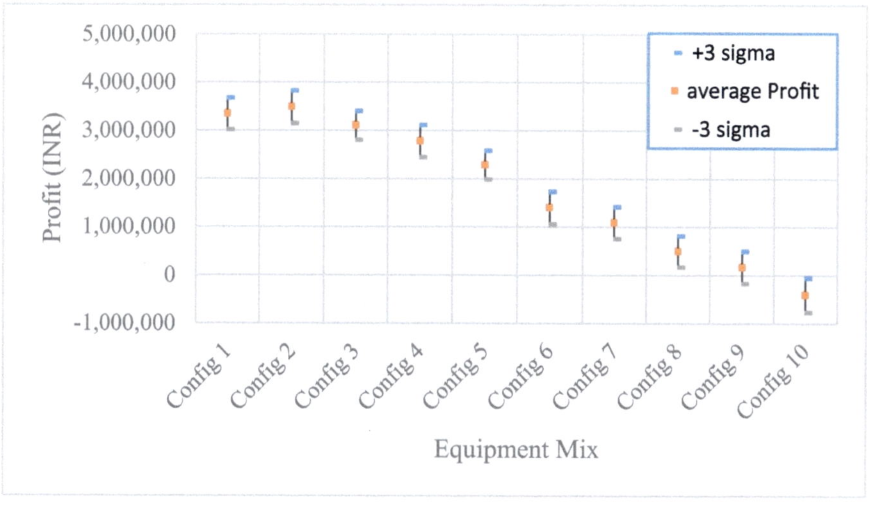

Fig. 5. Expected profit for different CHC configurations

To dig deeper, and calculate the probability of profit and loss, 100 simulations were run and results plotted (Fig. 6) to illustrate this special case. It shows that the probability of profit is more than the loss. Similar analysis can be conducted for other configurations.

Such analyses can help entrepreneurs choose the best configuration of a CHC, based on the initial investment needed, expected profits, and risk appetite.

It also helps in evaluating the impacts of various factors, those which entrepreneur can control (e.g. Equipment to buy, hiring rent, schedule, etc.) those that can be influenced (e.g. number of farmers asking for equipment i.e. customer base, farmers' patience, etc.) and those which are totally out of control (e.g. cropping pattern, season or time window of operation, fuel price, etc.). All these factors impact the profitability of the CHC. As an example, Fig. 7 illustrates how even a 10% change (decrease or increase) in equipment rent affects profit drastically. This small change in just one factor can make or break the CHC.

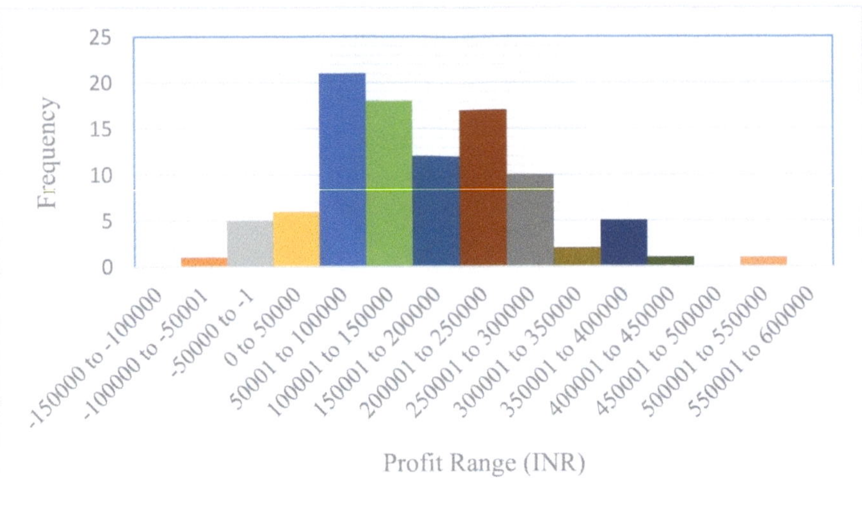

Fig. 6. Histogram for expected profit of configuration 9

Fig. 7. Impact of change in rent on expected profit of configuration 9

4   Conclusions

In this paper, we have presented an approach to simulate the operations of a CHC, and thereby estimate profitability, given a particular equipment selection. The simulation model is developed using a combination of Agent-Based Modeling and process modeling or Discrete Event Simulation approaches. We have indicated how the model can be used to make equipment selection choices when establishing a CHC in a particular location. With these results, it can be concluded that simulation, and specifically Agent Based – Discrete Event modeling can be useful to model Custom Hiring or service center operations. Particularly, farmers' behavior in different seasons, for different crops with different landholding and patience in getting the machines can be represented. The results obtained show that even with fixed input parameters, there is a lot of difference in net cumulative income. Such approach will help Original Equipment Manufacturers (OEMs) suggest probable customers appropriate number of machines to buy. This may also help banks and other financial institutions to make conscious decisions while disbursing the loans.

The approach can also be used to make other decisions for CHC operations. For example, the effect of prioritizing requests instead of serving them on a first come first served basis, choices between single purpose and multi-purpose equipment, etc., can be studied.

The current model assumes rudimentary behavior of the CHC and farmers that rent equipment from the CHC. However, the approach is extensible to incorporate complex behavior including effects due to weather changes, effect of interaction and word of mouth among farmers, effect of machinery breakdowns and other chance events, etc.

The benefits of CHCs as means to extend the benefits of mechanization to small and marginal farmers has been well accepted. However, there is a need to provide better guidance to entrepreneurs in establishing and operating these CHCs to ensure that they are profitable and sustainable. Through the modeling approach presented in this paper, and the extensions to this approach, we hope a complete set of tools is available for providing such guidance to entrepreneurs and policy makers.

References

1. Hiremath, G.M., Lokesh, G.B., Maraddi, G.N., Patil, S.: Accessibility of farm machinery services - CHSCs for small and marginal farmers. Int. J. Manag. Soc. Sci. **3**(2), 897–907 (2015)
2. Srinivasarao, C., Dixit S., Srinivas, I., Sanjeeva Reddy, B., Adake, R.V., Borkar S.: Operationalization of custom hiring centres on farm implements in hundred villages in India, Central Research Institute for Dryland Agriculture, Hyderabad, Andhra Pradesh (2013). 151 p
3. Sidhu, R.R., Vatta, K.: Improving economic viability of farming: a study of Cooperative Agro Machinery Service Centres in Punjab. Agric. Econ. Res. Rev. **25**, 427–434 (2012)
4. Chahal, S.S., Kataria, P., Abbott, S., Gill, B.S.: Role of cooperatives in institutionalization of custom hiring services in Punjab. Agric. Econ. Res. Rev. **27**, 103–110 (2014)
5. Kamboj, P., Khurana, R., Dixit, A.: Farm machinery services provided by selected cooperative societies. Agric. Eng. Int. CIGR J. **14**(4), 123–133 (2012)
6. Paman, U., Bahri, S., Asrol, A.: Custom hiring services of power thresher for small-farm rice threshing in Kampar Regency, Indonesia. Int. J. Adv. Sci. Eng. Inf. Technol. **4**(4), 274–277 (2014)
7. Schuler, R., Frank, G.: Estimating agricultural field machinery costs. learningstore.uwex.edu/pdf/A3510.pdf. Accessed 21 Sept 2015
8. Ag Decision Maker: Crops - Machinery. http://www.extension.iastate.edu/agdm/cdmachinery.html. Accessed 20 May 2015
9. Singh, K., Mehta, C.R.: Decision support system for estimating operating costs and break-even units of farm machinery. Agric. Mech. Asia Afr. Lat. Am. **46**(1), 35–42 (2015)

10. Maisenbacher, S., Weidmann, D., Kasperek, D., Omer, M.: Applicability of agent-based modeling for supporting product-service system development. Procedia CIRP **16**, 356–361 (2014)
11. Eid, M.S.: Optimization of service-equipment fleet by simulation. In: Winter Simulation Conference (1981)
12. Cook, T.: Evaluation of optimal fleet type and size for community transportation systems (2012)

Modelling the Efficacy of Assurance Strategies for Better Integration, Interoperability and Information Assurance in Family-of-System-of-Systems Portfolios

Keith Joiner[1], Mahmoud Efatmaneshnik[2(✉)], and Malcolm Tutty[3]

[1] School of Engineering and Information Technology, Capability Systems Centre, University of New South Wales – Canberra, Australian Defence Force Academy, Campbell, ACT 2612, Australia
k.joiner@adfa.edu.au
[2] Capability Systems Centre, University of New South Wales – Canberra, Australian Defence Force Academy, Campbell, ACT 2612, Australia
m.efatmaneshnik@adfa.edu.au
[3] Air Power Development Centre, Royal Australian Air Force, Fairbairn Offices, Department of Defence, Canberra, ACT 2600, Australia
malcolm.tutty@defence.gov.au

Abstract. Military systems, and more broadly society's, are increasingly complex and interconnected enabling hitherto only dreamed of capabilities and yet also humanity's forays into wholesale malicious cyber-warfare. Loosely coupled families-of-systems of systems cooperate and evolve sporadically when using linear lifecycles and project-by-project development, defying capability control and assurance at that mesa-level. The U.S. Defense has evolved systematic ways for their families-of-systems to be progressively more integrated, interoperable and information assured and this is dramatically differentiating its capability assurance from its allies. This paper reports new Markovian testability modelling comparing the abstract efficacy of assurance experimentation and testing strategies employed by Australia Defence against the new U.S. strategies that are now able to quantitatively illustrate the widening gap between these allies. The modelling technique has potential to tailor Australian plans to keep pace with its ally and in modelling civilian families-of-system-of-systems in transportation, energy healthcare and the like.

1 Introduction

Research in the U.S. has documented the increased complexity and interconnectedness of systems [1, 2], such that most commercial and public sector organizations operate multiple complex, interdependent and adaptive systems. Such complexity and interconnectedness has been well predicted [3, 4] and there are numerous strategies to deal better with them including: scientific experimentation and test methods [5–7] software development and test methods [8, 9], preview or early testing [10, 11], cyber-resilience

© Springer Nature Switzerland AG 2019
M. A. Cardin et al. (Eds.): CSD&M 2018, AISC 878, pp. 25–36, 2019.
https://doi.org/10.1007/978-3-030-02886-2_3

methods [12, 13], complex systems governance [14, 15], and for intelligent/autonomous systems design and test [16]. All these innovative methods and approaches are fundamentally aggregated in, or enabled by, the systematic experimentation and test assurance methods compared in this work because these strategies are at a meta-level of systemically managing capability systems.

Joiner and Tutty [17] examined how the U.S. Department of Defense (DoD) had since 2009 undertaken six interrelated initiatives to significantly affect more integrated, interoperable and information (I3) assured Family of Systems (FoS). FoS is defined by federations or coalitions of intergenerational and complementary Systems of Systems (SoS) often from different countries and on multiple types of missions that have complexity and adaption beyond what was envisioned in their design and sustainment [18]. Such initiatives are both to cope with such complexity and interconnectedness and to exploit it for information dominance which is key to fifth generation warfare [19]. These initiatives leverage or enable the many smaller initiatives outlined earlier into the six broad reform themes and hence they harness innovation in multiple disciplines. Joiner and Tutty [17] also qualitatively compared how the U.S. DoD approach differs to the more contemporary approach taken by the Australian DoD so far and then concluded there was a significantly widening gap in the efficacy of these country's forces, especially for allies that seek close alignment. This work models the two DoD I3 assurance strategies and illustrates quantitatively how the two approaches to DoD I3 assurance, even if operating similar systems can, lead quite different operational utilities.

Efatmaneshnik et al. [20] analyzed the effect of modularity on test efficiency and cost, and in more general terms the relationship between systems testability and system architecture. They presented a Markovian model of system testing that captures the rework and process repeats. This paper uses a Markovian process model to evaluate the efficacy of the two assurance strategies. One of the key points of the testability research by [20] is the importance of repeating tests and how the most-efficient number of repeats at each level in sub-system verification is usually not intuitive and should be optimized as part of early design. For the FoS, there are both *in-service* SoSs undergoing incremental sustainment and other evolution, as well as *developing* SoS that will regularly replace obsolete SoS usually with technological and capability enhancements. Herein is a generational tension worthy of the term *family*, where the FoS in part does not seek to control an SoS but rather to evolve through its incorporation, and that one of the existential threats to the developing SoS and the future FoS in which it resides, is the obsolescence of the FoS against new threats when compared to the developing SoS.

2   Contemporary Australian Approaches to I3 Assurance

This section will outline what the average or typical Australian DoD approach to I3 assurance is using cross-domain reviews like those of the Australian Senate and Australian National Audit Office (ANAO) [21]. It then proposes a simplistic model of the average level of I3 assurance testing, conscious of that fact that there is an enormous diversity of approaches between operational elements, their sustainment offices and especially among acquisition projects. One of the authors reviewed every project submission to Government for four and half years and worked on the ANAO and Senate reviews and can attest that the problem is not just where the average is, but the disappointing lack of consistency from excellent through to very ordinary, especially if the lens is narrowed to, '*what objective test activities will occur from reasonably possible modelling or OTS components before contract?*' This lens sees phrases like 'rushing to contract', 'project over-optimism', 'conspiracy of optimism' and decisions by committees rather than test-based evidence.

2.1   Test and Evaluation Supporting Systems Engineering and Program Management

The Australian DoD approach to I2 assurance focuses on operational exercises to do the team exercise for the adults of the FoS. Child SoSs, in the main, cannot and do not participate in operational exercises due to a lack of maturity. Unlike the U.S. DoD, on average, no additional integration or experimental exercises exist to allow child SoSs to play with adults in a controlled environment.

The child SoSs each go through an assurance program (i.e., school) envisioned and planned during the requirements and early contract phase and implemented quite late in development during acceptance and operational testing as the systems are built, assembled and delivered. I2 assurance before contract is focused on reviewing the written requirements against future operating and integrating concepts. Checks that the child SoS can be integral and trusted in the FoS are based largely on describing the SoS boundaries and then testing to those boundaries much later based on interface control documents and standards.

In essence, most Australian DoD projects follow the path of least resistance from weak acquisition policies and therefore are paper-based for all of the definitional phase and then heavily outsource the majority of verification testing of sub-systems, systems and SoS to the developmental contractor. The developmental contractor usually has little access to operational assets or systems from the FoS and no real incentive under the contract to get that representation [21]. The average result is a lot of surprises at the operational testing, increasingly because the FoS has adapted over the development period and requirements for the child SoS have been in comparative stasis [17]. The frozen requirements can often be characterized as an alternate reality based on the fairness of original specifications and operational concepts, most of which were never subject to activity testing. The use of modelling and simulation, prototypes and distributed federated T&E can obviously have significant positive effects on child SoSs, however each project during scoping has to decide without real contractor knowledge what risks exist and what resources to allocate to such modelling and simulation-based T&E. If modelling occurs, it is usually the development contractor's decision to de-risk their portion of the life-cycle, help use it to simplify verification tests and eventually to deliver it in support of operator training at the transition. The test agencies in Australia and the acquisition policies that support them are unfortunately not set up like the U.S. DoD to accredit such models [22] or to evaluate capabilities for objective independence at the '*into contract*' and '*into production*' milestones [14, 15].

2.2   Current Australian I3 Assurance Model

The I3 assurance test model for the Australian DoD is shown in Fig. 1 and has a number of limitations. Most significantly, the model does not directly include the adult SoSs that constitute the FoS and instead focuses on the development of each child SoS. This limitation mainly effects the fidelity of the later U.S. DoD model since the U.S. DoD uses experimentation exercises to augment operational exercises to assure the I3 of their FoS, meaning it can be more actively developed and prioritised. There is some overlaid effects on the Australian model to account for this limitation, mainly through the Assured Utility (AU) of the SoS after each test, where it is kept low to reflect the uncertainty of the FoS against which utility should be judged.

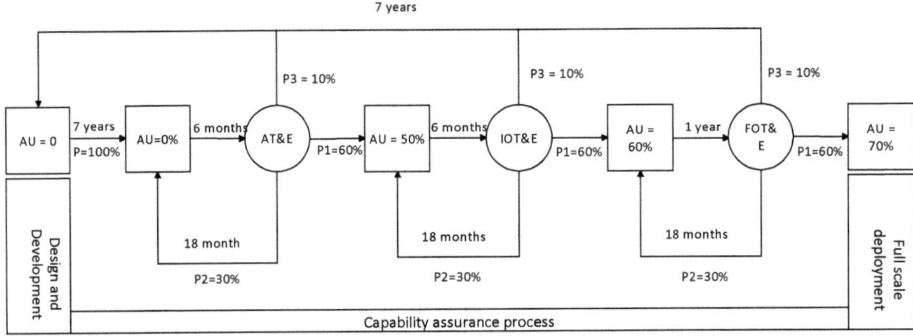

Fig. 1.   I3 assurance test model for the Australian DoD with illustrative figures.

The Australian DoD I3 assurance model has an average SoS project life of 10 years and begins with zero utility for the first 7 years while the needs and requirements phases are conducted and the design occurs. If this zero utility seems harsh, it is based on the average likelihood of a valid test report existing with representative users before acceptance T&E (AT&E). The model then has three test phases each an average of a year-long, being: AT&E, Initial Operational T&E (IOT&E) and Final OT&E (FOT&E). Each test involves a probability of passing (P1), failing in a minor way (P2) such as a software recoding with an 18 month reset and failing majorly (P3) and going back to the contract stage (7 year reset). Each test improves the AU by an estimated amount. The AU values reflect a skeptical user about the entire process, including whether the operational concept and specification was right in the first place, has been kept up-to-date and to reiterate from earlier, that the FoS needs of that developing SoS are well understood and current (i.e., school remains appropriate). For example a hard utility to judge in the Australian context for nearly all capabilities is, *'has the FoS been subjected to representative cybersecurity OT&E so that the SoS's attack surface is fully known and tested?'* [12]. The results of running simulations with this model are provided later. Importantly for potential in tailoring the model to specific circumstances, all variables can be adjusted.

3   New U.S. Defense Approach to I3 Assurance

Underpinning the U.S. DoD federated systems integration laboratories (SILs) is the concept and protocols of their live-virtual-constructive (LVC) simulation capabilities that are required for all capabilities, must be accredited [22], and which significantly de-risk capabilities for aspects like usability. However, in the particular context of this review, the federated SILs and LVC simulation capability de-risk integration of the child SoS with their FoS. A diagrammatic representation of these I3 assurance activities is in Fig. 2.

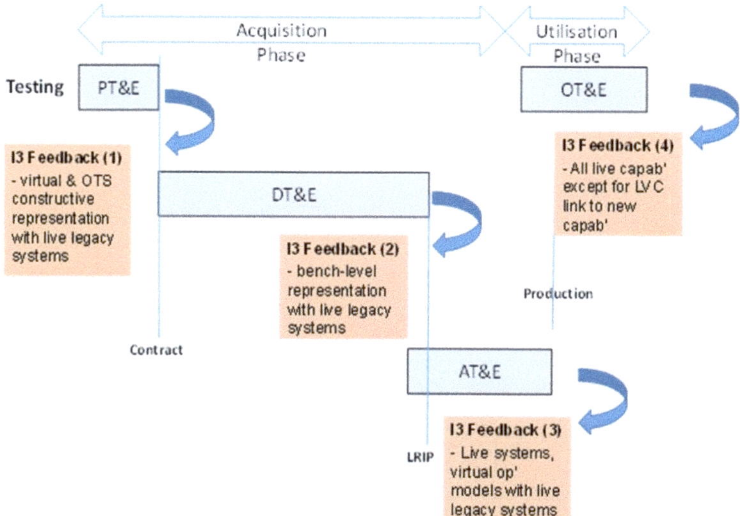

Fig. 2. I3 assurance test model for the U.S. DoD using Australian T&E descriptors. PT&E = preview T&E; DT&E = developmental T&E; AT&E = acceptance T&E; OT&E = operational T&E; LRIP = low-rate initial production. Bench-level means key systems may be hardware-in-the-loop and key software modules may be the standard of a software integration laboratory.

One of the most powerful effects of an I3 assurance test regime like that shown in Fig. 2 is to bring together operational testers, who manage in-service testing, with developmental testers, who are often contractors, to achieve a shared test on I3 assurance. These two cultures then have to agree: (1) system performance metrics when embedded in the FoS, (2) common test planning methodologies, which is usually the efficient high-throughput testing or design-of-experiments methods delivered under a different initiative [6]; and (3) when they find impediments to integration or cyber-vulnerabilities, characterise these and agree to recommendations to fix. Anyone who has managed the progression of capability from contractor development through into service, should recognise the significant cultural benefits, risk-reduction, innovation and trade-off from such combined test efforts.

3.1 U.S. Defense I3 Assurance Model

The I3 assurance test model for the U.S. DoD is shown in Fig. 3 that has the additional test pieces (relative to Australian model) in the design and development phases. The U.S. DoD model has a number of limitations. Like the Australian model it does not directly include the adult SoSs that constitute the FoS and instead focuses on the development of each child SoS. This limitation mainly affects the U.S. DoD model since the U.S. DoD uses experimentation exercises to augment operational exercises to assure the I3 of their FoS. The U.S. DoD model partly accounts for this limitation, by improved probabilities of achieving the Assured Utility (AU) at the latter parts of the I3 assurance model (i.e., AT&E to FOT&E) when compared to the Australian model; however, the actual AUs have been kept the same between models where this is probably pessimistic assumption.

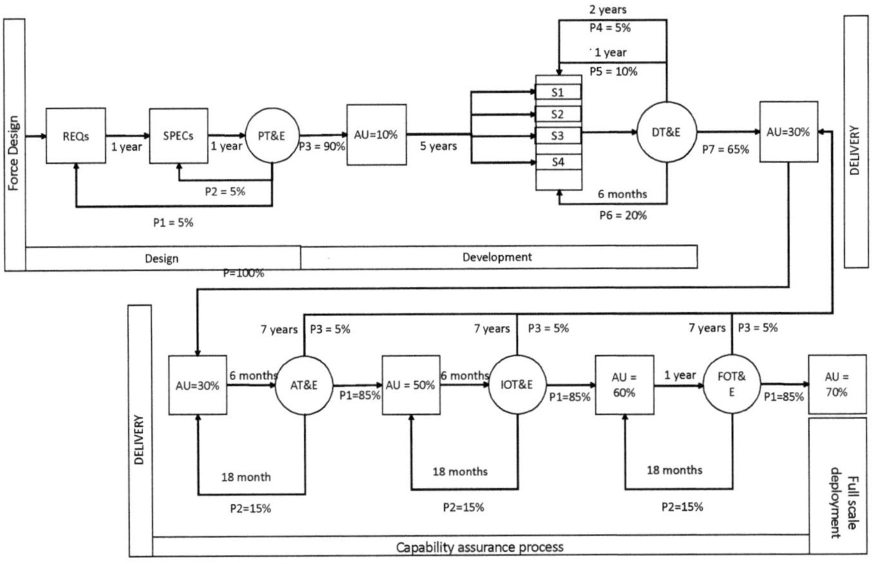

Fig. 3. I3 assurance test model for the U.S. DoD that has pre-delivery and post-delivery assurance processes.

The U.S. DoD I3 assurance model has the same average SoS project life of 10 years and also begins with zero utility but only until Preview T&E (PT&E) after 2 years. The PT&E is pre-contract and is a de-risk by users on a mixture of virtual models and OTS systems to determine a modest amount of utility (10%), check I3 with distributed FoS and confirm operational concepts and high-level requirements. This is followed in developmental T&E by key systems coming on-line to the user model during development so as to update usability, I3 to the FoS and cyber-resilience overall—giving another estimated 20 percent of early utility. Because of these early evaluations, the probabilities of rework are set lower in US DoD I3 model. The results of running simulations with this model are provided later. Importantly for potential in tailoring the model to specific circumstances, all variables can be adjusted.

4   Computational I3 Assurance Comparison

In this section the models in the two I3 assurance models presented in Figs. 2 and 3 were used as stochastic Markov processes that can be statistically studied and compared. The depth and breadth of the models are suitable for a preliminary analysis of the two assurance schemes. However, the model is extendable both in terms of breadth (i.e. the incorporation of more influencing variables such as test system types) and depth (i.e. the consideration of information that validate the probabilities and times). The models presented in Figs. 2 and 3 are indeed absorbing Markovian processes, that can be either solved using classic Markov chain formulations or alternatively by Monte Carlo Simulation. We chose Monte Carlo Simulation for simplicity and the greater insight it provides.

Monte Carlo simulation of the models were conducted in MATLAB environment. A sample size of *100,000* random scenarios are generated for each model. Each scenario corresponds to a possible path in the assurance models, and the number of times a particular path appears is representative of its occurrence probability.

Figure 4(a) shows a classification of the simulation results for Australian model (a) and US model (b). These plots are intended to provide the general overview of the

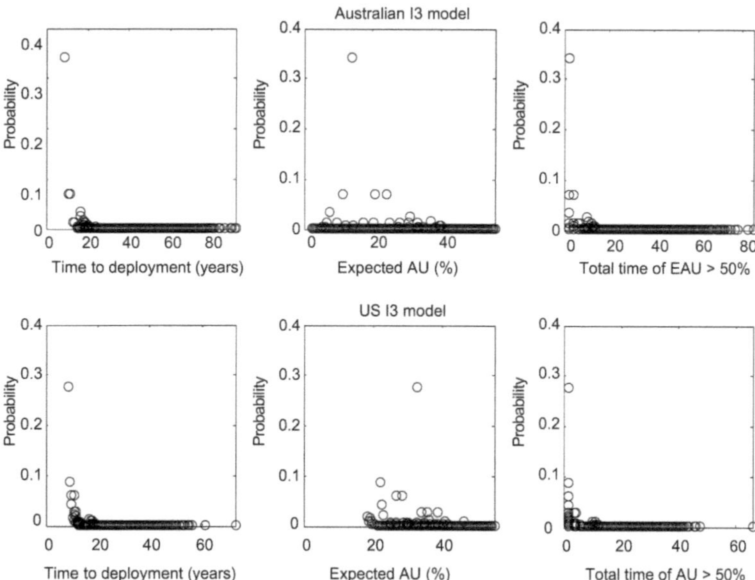

Fig. 4. The top and bottom graphs show the probabilities and three variables of each possible scenario resulting from the probabilities shown respectively in Australian and US I3 assurance models. The three variables are, Total time, Expected AU over development and T&E periods, and total time with AU of more than 50% during T&E.

scenarios land scape. Figure 5 shows the cumulative probabilities of some variables used for estimation of confidence intervals in Table 1.

Table 1. The overall comparison of Australian and US assurance models.

Parameter	Australian model	US model
Mean time to deployment	15.1	12.1
Expected AU development and assurance time	20.81%	30.69%
Time to deployment (years) - 90% confidence interval	[9, 26]	[9, 18.5]
AU over development and assurance time (%) - 90% confidence interval	[2.77%, 37.87%]	[18.36%, 38.69%]
Total time with AU > 50% (years) - 90% confidence interval	[2, 15.5]	[2, 9.5]

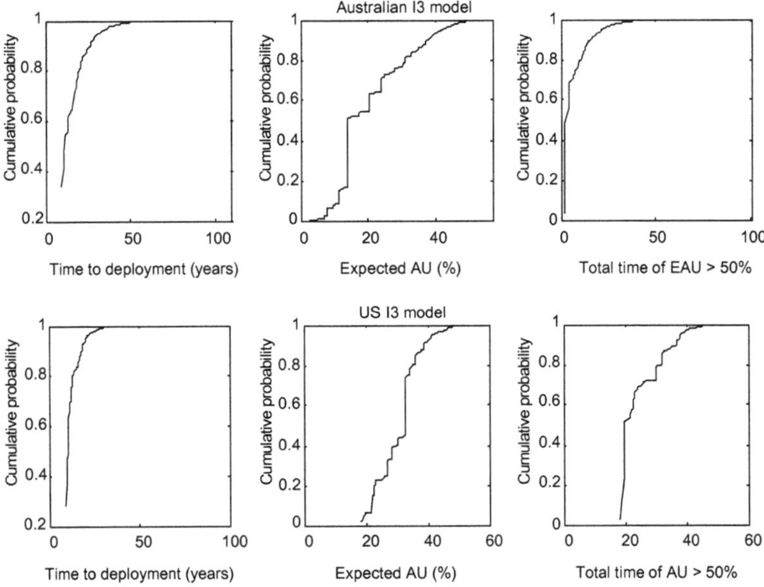

Fig. 5. The cumulative probabilities of the same variables as in Fig. 4.

Three variables were chosen to show the qualities of each scenario: Total time to achieve set AU for the characterized SoS, expected AU over development and T&E periods, and total time with AU of more than 50% during T&E. The probabilities are calculated based on the number of repeats of the same scenario divided by the total number of generated samples. For both models, we can see one solution with the probability of around 30%. This solution relates to the path where no rework loop happens (lowest time), and naturally has a lower probability for the U.S. model because it has three more T&E related loops in the development phase.

For overall comparison of the model we chose these key parameters:

(1) Mean time to deployment: reflects the expected time resulted from using either T&E model over many projects. It is calculated based on the mean of the solutions times or the weighted average of scenarios scaled by their probabilities. The average time for Australian model is about 15.1 years and for US model is about 12.1 years.
(2) Expected AU: is representative of average AU over the life of the acquisition project. The average utility for Australian model is about 20.81% and for US model about 30.6%.
(3) 90% confidence interval for time to deployment.
(4) 90% confidence interval of AU over the project life.
(5) 90% confidence interval of total time with AU > 50%.

The comparisons of the two models based on the above parameters is summarized in Table 1. The table shows that the U.S. model has potential for substantially better outcomes with regards to all of the above stochastic parameters. Note that this is the case despite the fact that the assumed delivery time in Australian model was 7 years. However, for the US model, because of the three additional loops, the expected delivery time is about 7.6 years with a minimum 7 years, and a standard distribution of 0.9 years.

4.1 Limitations

One of the key limitations of a Markov model is that testing of the unit is assumed to be simplistically as a memory-less process in the sense that a new test is performed based on the state of the prior test only, regardless of all the previous tests. In some respects in a FoS in a large organisation over so many years the bureaucracy helps give weight to this assumption. As projects work faster in their development and more closely work with their FoS then Markov assumptions are less likely to be realistic. Indeed a FoS that might themselves be time-critically developed and tested for a conflict is likely to have better awareness of their criticalities and stay more closely aligned to child SoSs. The project domain of Counter Improvised Explosive Device (IED) may be an example of an area where this limitation applies.

Other key limitations come from setting a single project I3 assurance model of an average duration to testing for each country, when the projects delivering SoS are so diverse, the programs managing in-service FoS are so diverse, and finally, policy only really guides such management and thus there is likely to be many worse and better exceptions.

5 Conclusion and Future Work

Markovian modelling has been developed to quantitatively represent and then compare the different policy-based average test strategies in the U.S. and Australian DoDs for assurance of integration, interoperability and information between the SoS that they operate in their FoSs. Despite significant limitations in such probabilistic mesa-level comparison, the modelling found quantitative estimates confirm qualitative work by

Joiner and Tutty [17] as to the significant effect in the U.S. DoD of the early testing using federated SILs and modelling and simulation. The extra *observability points* between developing SoS and the in-service FoS, while not giving *controllability*, improves the consistency of I3 in the *schooling* of each SoS and the *team exercises* of the FoS. Empirical work needs to be done to confirm estimates at each test phase, and indeed to better represent differing classes of SoSs, but the work shows that the early test strategies of the U.S. deliver an expected total project SoS development time of three years sooner on average than the Australian approach.

Most importantly though for an assurance test strategy, it is not the averages that convey the true benefit of the extra assurance investment but rather the confidence limits around the total project SoS time. The 90% confidence limits for the best project SoS times are the same for both strategies, such that some projects delivering SoS will do as well in either countries' assurance regime. However, the 90% confidence limits for the worst project SoS times are 8.5 years longer for Australia (Table 1), such that some projects delivering SoS will do substantially worse in Australia or in many cases get cancelled trying (i.e., 26 years). This works supports empirical and policy work in both countries as to the substantial benefits of early de-risk testing and technical maturation before contract [10, 11].

The most significant limitation to work on in the models is better represent the in-service FoS or pool of SoSs being integrated and information assured for a common purpose. The model as such is still project-centric or SoS by SoS development focused and needs to better represent the sustainment and operational management challenge of the whole FoS. For example, since the FoS of each countries' DoD consist of multiple SoS, progressively the Australian DoD will develop slower and with less I3 than the U.S. DoD and the model cannot yet show family cohesiveness or degradation over time. Also, the effects of the spread in possible project SoS times on force design is likely to be very disruptive, since the predicted force composition in ten years' time would have a significant portion failing to arrive for another decade. The model is yet to replicate the synergistic benefit of such early de-risk I3 assurance testing to work out which of the developing SoS are more or less likely to achieve utility and to plan accordingly for the *family* to cope with more or less *schooling* for all the *children*.

There is significant scope to tailoring this research work to different FoS used by each military, such as analyzing separately each of Australia's 40 programs ranging from soldier combat systems to submarines. This could enable prioritizing greater I3 assurance testing for those capabilities with shorter readiness times, greater allied interoperability needs, or substantially different technology advancement rates or development timelines. A scale or tiers of I3 assurance test approaches might be possible. There is also scope to apply this research to Civilian fields with analogous FoS, such as places where critically continuous operations of a diverse nature require replacement SoS to be progressively added by accretion in ways not entirely designed. Examples include some commercial markets, healthcare systems, energy systems, some ICT networks, transportation networks and so forth.

Acknowledgement. Authors are gracious for rich and helpful dissuasions with CMDR Kate Miller, whose insights facilitated shaping this paper.

References

1. Jenner, S.: Why do projects 'fail' and more to the point what can we do about it? The case for disciplined, 'fast and frugal' decision-making. Management **45**, 6–19 (2015)
2. Hecht, M.: Verification of software intensive system reliability and availability through testing and modeling. ITEA J. **36**, 304–312 (2015)
3. Courtney, J., Merali, Y., Paradice, D., et al.: On the study of complexity in information systems. Int. J. Inf. Technol. Syst. Approach **1**(1), 37–48 (2008)
4. Flyvbjerg, B., Budzier, A.: Why your IT project may be riskier than you think. Harv. Bus. Rev. 24–27 (2011)
5. Ahner, D.K.: Better buying power, developmental testing, and scientific test and analysis techniques. ITEA J. **37**, 286–290 (2016)
6. Joiner, K.F.: Six-sigma reform and education in Australian Defence: lessons-learned give rigour and efficiency to ordnance, aircraft and ship testing. In: 7th International Conference on Lean Six Sigma, Dubai, United Arab Emirates (2018)
7. Murphy, T., Leiby, L.D., Glaeser, K., et al.: How scientific test and analysis techniques can assist the chief developmental tester. ITEA J. **36**, 96–101 (2015)
8. Cofer, D.: Taming the complexity beast. ITEA J. **36**, 313–318 (2015)
9. Kuhn, D.R., Kacker, R.N., Feldman, L., et al.: Combinatorial testing for cybersecurity and reliability (2016)
10. Copeland, E.J., Holzer, T.H., Eveleigh, T.J., et al.: The effects of system prototype demonstrations on weapon systems. In: Defense Acquisition Univ FT Belvoir VA (2015)
11. Joiner, K.F.: How new test and evaluation policy is being used to de-risk project approvals through preview T&E. ITEA J. **36**, 288–297 (2015)
12. Joiner, K.F.: How Australia can catch up to US cyber resilience by understanding that cyber survivability test and evaluation drives defense investment. Inf. Secur. J. Glob. Perspect. **26**, 74–84 (2017)
13. Mead, N.R., Woody, C.: Cyber Security Engineering: A Practical Approach for Systems and Software Assurance. Addison-Wesley Professional, Boston (2016)
14. Joiner, K.F.: Implementing the defence first principles review: two key opportunities to achieve best practice in capability development. In: Strategic Insights: Australian Strategic Policy Institute (2015). www.aspi.org.au
15. Keating, C.B., Katina, P.F., Joiner, K.F., et al.: A method for identification, representation, and assessment of complex system pathologies in acquisition programs. In: 15th Annual Acquisition Research Symposium, Graduate School of Business & Public Policy at the Naval Postgraduate School, Monterey, California (2018)
16. Association IS: Standard for Fail-Safe Design of Autonomous and Semi-Autonomous Systems P7009 (under development) (2018)
17. Joiner, K.F., Tutty, M.G.: A tale of two allied defence departments: new assurance initiatives for managing increasing system complexity, interconnectedness and vulnerability. Aust. J. Multi-Discipl. Eng. **14**, 4–25 (2018)
18. Tutty, M.G.: The profession of arms in the information age: operational joint fires capability preparedness in a small-world. University of South Australia (2016)
19. Layton, P.: Fifth-generation air warfare. Aust. Def. Force J. **204**, 23–32 (2018)

20. Efatmaneshnik, M., Shoval, S., Ryan, M.: A framework for testability analysis from system architecture perspective. In: INCOSE Annual Symposium, Washington, D.C., 18–22 July 2018 (2018)
21. Australian Senate and Australian National Audit Office (ANAO): Test and Evaluation of Major Defence Equipment Acquisitions (2015)
22. Elele, J.N., Hall, D.H., Davis, M.E., et al.: M&S requirements and VV&A requirements: what's the relationship? ITEA J. **37**, 333–341 (2016)

Multi-class Fleet Sizing and Mobility on Demand Service

Malika Meghjani[1(✉)], Scott Drew Pendleton[2,3], Katarzyna Anna Marczuk[2,3], You Hong Eng[1], Xiaotong Shen[1], Marcelo H. Ang Jr.[2], and Daniela Rus[4]

[1] Singapore-MIT Alliance for Research and Technology, Singapore, Singapore
`{malika,youhong,xiaotong}@smart.mit.edu`
[2] National University of Singapore, Singapore, Singapore
`{scott.pendleton01,katarzyna}@u.nus.edu, mpeangh@nus.edu.sg`
[3] nuTonomy, Singapore, Singapore
[4] Massachusetts Institute of Technology, Cambridge, USA
`rus@csail.mit.edu`

Abstract. This paper addresses multi-class fleet sizing and vehicle assignment problem where we aim to provide Autonomous Mobility-on-Demand (AMoD) service using a fleet of heterogeneous vehicles. We present a chain of transportation with three classes of autonomous vehicles including cars, buggies and scooters. Each class of vehicle can access a subset of the network, such that, there are some links exclusive for that particular class. Our fleet management system then assigns available vehicles to trips based on the travel time for passenger pick-up and drop-off, their queue time and accessibility of the road network by the vehicle. Each assignment may consist of a set of vehicles allocated for one trip that is composed of multiple-legs served by different vehicles. For example, first mile pick-up by a scooter, middle mile on a car and last-mile trip on a buggy. We apply a genetic algorithm for heterogeneous fleet sizing and propose a hierarchical structure for travel time optimal assignment of the multi-class autonomous vehicles to passengers. We validated our approach with a range of heterogeneous fleet sizes constrained on the given budget. Our approach is more time efficient than taking a ride on a single-class autonomous vehicle for middle mile plus walking during the first and the last miles. Hence, we provide the convenience of autonomously covering the entire journey using multi-class vehicles with no additional travel or transit delays compared to single-class.

1 Introduction

Autonomous Mobility on Demand (AMoD) systems provide demand-responsive transportation services using self-driving vehicles. These vehicles generally cover partial to complete journeys of the passengers, depending upon the vehicle accessibility to the road network. Our goal is to specifically provide complete coverage from first mile to last mile using multi-class autonomous vehicles. Our fleet of multi-class vehicles, shown in Fig. 1, range from slow-drive, easily navigable,

© Springer Nature Switzerland AG 2019
M. A. Cardin et al. (Eds.): CSD&M 2018, AISC 878, pp. 37–49, 2019.
https://doi.org/10.1007/978-3-030-02886-2_4

personal mobility scooters for individuals, to medium speed autonomous buggies and faster speed self-driving cars with larger capacity and accessibility to larger lanes and main streets.

Fig. 1. SMART's multi-class autonomous vehicles: road car (upper left), buggy (lower left), scooter (right).

The specifications for each of the autonomous vehicles is given in Table 1. These specifications are also represented by a star diagram for easier comparison in Fig. 2, where larger values indicate better performance. The car is most ideal for travel along the road network, as it has the highest speed and is capable of longest range. However, the scooter would be more ideal for narrow passageways and crowded pedestrian environments. It is the cheapest platform, most efficient in terms of weight, most maneuverable (smallest turning radius), and small enough to fit inside building hallways and on small sidewalks. The buggies are most well suited for large pedestrian areas, such as parks, plazas, airport terminals and hospitals.

By using all three classes in combination, a greater accessibility and service coverage can be achieved such that the users can be taken not just between building pick-up and drop-off points, but even from the room of one building to a specific room in another building several kilometers away. Our proposed multi-class mobility on demand system is best applicable to elderly or disabled people who might need assisted modes of navigation. In addition, it is also very useful for passengers and/or goods transportation at facilities with long transits (e.g. airports and hospitals) or under bad weather conditions.

An ideal application scenario for multi-class AMoD service is a campus network where the shuttle buses provide transportation between the major stops

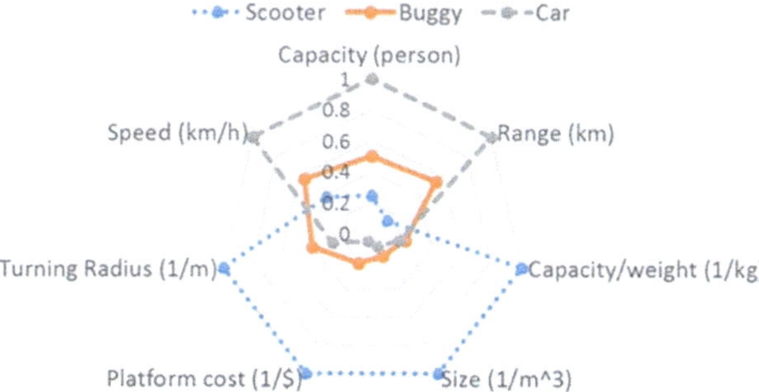

Fig. 2. Comparison of vehicle classes based on normalized values extracted from Table 1. Higher values are more ideal.

Table 1. Self-driving vehicle specifications

	Car	Buggy	Scooter
Length (mm)	3885	2525	930
Width (mm)	1515	1200	485
Height (mm)	1750	1890	2100
Empty Weight (kg)	1200	500	56
Seating Capacity (persons)	4	2	1
Operating Speed (km/h)	19	10.8	7.2
Range (km)	150	80	20
Turning Radius (m)	4.5	3.0	1.2
Platform Retail Cost (USD)	29125	7499	1599

for the middle mile and the passengers are required to walk to cover their first and last mile. Our goal is to analyze this campus scenario where the transit points are fixed and the shuttle bus schedules are almost certain. Given this scenario, we first solve the multi-class fleet sizing problem to decide on the number of vehicles in each class. We then propose a multi-class vehicle assignment algorithm to allocate combinations of road cars, buggies, and personal mobility scooters to the passengers. Our solution has the travel time efficiency better than single-class assignment at the convenience of multi-class door-to-door service. We provide a thorough case study of the campus scenario in simulation. Our proposed method is generalizable to other hierarchical transportation models involving private, shared and public transportation systems for inter and intra-city commutes. A proof of concept implementation for multi-class mobility on demand system with real autonomous vehicles (one in each category) was also illustrated in our previous work [1]. The contributions of this paper are as given below.

- A multi-class fleet sizing algorithm for deciding the number of vehicles required in each class, given an estimated demand size and expected budget.
- A generic and scalable multi-class assignment algorithm for providing door-to-door transportation.
- An empirical analysis in a real-world scenario given the fleet sizing and vehicle assignment algorithm.

Each of these contributions are highlighted in the following sections. Specifically, Sect. 3 outlines our general software architecture for multi-class management system, Sect. 4 provides algorithms for heterogeneous fleet sizing and vehicle assignment and Sect. 5 presents our simulation results.

2   Related Work

Research in the field of autonomous vehicles has matured in the past few years. Recent studies have also been conducted to model and anticipate the social impact of implementing AMoD [2,3]. The case studies have shown that MoD systems would make a more affordable and convenient access to mobility compared to traditional transportation system characterized by extensive private vehicle ownership [3].

Mobile Internet technology has created the opportunity to enable dynamic and on-demand transportation services, i.e., through e-hailing applications. These services have a potential to provide societal and environmental benefits when designed adequately. The core of the real-time e-hailing concept is the development of algorithms for optimally matching vehicles (or drivers in traditional systems) and passengers [3–5]. We have seen a growing interest in the intelligent transportation literature to address the optimization issues in the dynamic assignment for autonomous mobility on demand systems. As of today, the number of specific contributions is still small [6].

A demand-responsive personalized service to passengers is introduced in [7]. In this paper, the passengers have flexibility to choose the service type (taxi, shared-taxi and mini-bus) from a menu that is optimized in an assortment optimization framework. For operators, there is flexibility in terms of vehicle allocation to different service types, which implies that the vehicles are changing their class according to the demand pattern. This framework is tested on a small network. Similarly, [3,6] propose a personalized mobility solution with autonomous vehicles. In their work, the dispatcher assigns vehicles to trips (with single and multiple pick-ups) using Greedy and Bipartite assignment and with prediction for vehicle pre-positioning. The authors test their dispatcher in the simulation environment for the city of Singapore. However, they consider only one class of vehicle. Similarly, most of the theoretical approaches to the assignment problem usually attempt to solve it for single-class vehicles [8–10].

Autonomous driving on urban roads has seen tremendous efforts in the recent years. Google is at the forefront of these efforts, having tested its fleet of autonomous vehicles for more than 2 million miles, with expectation to soon launch a pilot MoD service project using 100 self-driving vehicles [11]. Autonomous systems have also been implemented in unstructured outdoor environments and urban pedestrian areas such as side walks and university campuses. A recent example is the Smart Wheelchair System developed at Lehigh University [12].

While the above mentioned research has shown promising results in their restricted cases, to our knowledge, this paper presents the first integrated solution that considers multiple types of vehicles for transport in both pedestrian and urban road environments. We first introduced the concept of multi-class AMoD in [13], where a self-driving buggy and a road car were used in combination. This concept was extended to three classes with the addition of a self-driving personal mobility scooter in [1]. In this paper, we present a novel hierarchical multi-class Fleet Management Service (FMS) algorithm tasked to assign large-scale multiple heterogeneous vehicles to each passenger request as deemed appropriate to the mission. We also address the multi-class fleet sizing problem which provides input to our multi-class FMS algorithm.

3   Software Overview

An overview of our AMoD system is shown in Fig. 3. The booking application within this system can be used by passengers to request a mobility service to travel from the *Pick-Up* location to the *Drop-Off* location. All the mobility requests are sent to a fleet management server which assigns Autonomous Vehicles (AVs) to passengers and generates unique verification codes corresponding to each request.

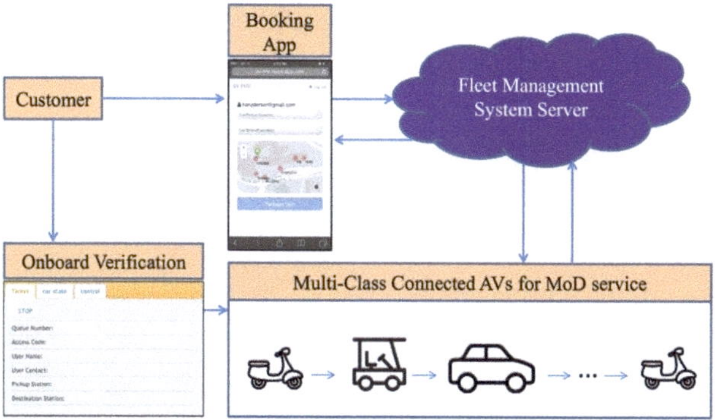

Fig. 3. Overview of our Autonomous Mobility-on-Demand (AMoD) system to provide connected transportation for mobility service [1].

Based on the traversability of the route, availability of the vehicle and the assignment cost, a suitable type of vehicle, such as a scooter, a buggy, a road car or their combination will be assigned to pick up the passenger from their origin and transit locations. Our assignment algorithm is specifically explained in Sect. 4. The assigned vehicle IDs are sent sequentially to the booking application to notify the passenger which vehicle(s) to board. Once the AV arrives at the *Pick-Up* location, the passenger can board the vehicle and key in their verification code to start the trip. The assigned AV will autonomously navigate to either the *Drop-Off* location to bring the passenger to the destination or a transfer station to allow the passenger continue the trip via another type of vehicle. Once the passenger reaches their destination or transit location, the FMS will assign the vehicle to a new user request for continuously providing the AMoD service.

4   Technical Approach

Given the network information, vehicle specifications, expected demand and the estimated budget, we first solve the fleet sizing problem. The multi-class fleet sizing problem for static demand (i.e., spatially distributed but not temporally) is similar to the bounded knapsack problem [14] where the total weight constraint of the knapsack is equivalent to our budget constraint and the items are vehicle classes with different weights and values. The weights of the items can be interpreted as the cost of the vehicle and the values of the items as the speed of the vehicle. However, we are interested in the multi-class fleet sizing problem for time-varying demand which is even more challenging than the aforementioned bounded knapsack optimization problem that is known to be NP-hard. Therefore, we focus on using approximation methods such as the genetic algorithm. The genetic algorithm [15] assesses different combinations of fleet sizes and evaluates the expected total travel time (sum of queue, assignment and travel time) per passenger using our proposed multi-class fleet assignment algorithm (Algorithm 1) as the fitness function. The components of total travel time are illustrated in Fig. 4.

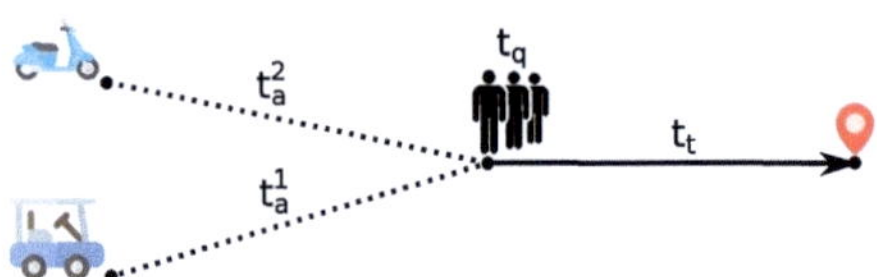

Fig. 4. Components of total travel time comprising queue time t_q, assignment time t_a and travel time to destination t_t

The multi-class fleet sizing algorithm begins with an initial guess for fleet size of each class of vehicle which needs to be provided as the input. This is used by the genetic algorithm to generate samples of fleet size combinations that satisfy the budget. These samples together form an initial population. It then creates a sequence of new combinations from the current generation of samples to create a new generation. Some of the combinations having the best fitness value (expected total travel time per passenger) from the current generation are retained and passed to the next generation. In addition, more samples of combinations are added to the next generation by either mutating or crossing over different combination sets. The sample combination corresponding to the overall best fitness value is considered as the final outcome for fleet sizing of each class of vehicle.

For the multi-class vehicle assignment problem, we propose a hierarchical approach. This process involves two stages of assignments corresponding to the first-last miles and middle mile of the journey. These assignments are performed sequentially for each passenger. We define fixed transit points to traverse from the first mile to the middle mile or from the middle mile to the last mile. We refer to these transit points as the main stations and the paths connecting them are only accessible by cars. Each of these main stations are part of a sub-network which is accessible by either scooters or buggies or both. The nodes in the sub-network are referred to as sub-stations. In our current work, we assume that all vehicles have a capacity of one passenger (i.e. no ride-sharing). An overview of our assignment algorithm is presented above in Algorithm 1.

Input: Locations of
- passengers $P = \{p_1, p_2, ..., p_q\}$
- scooters $S = \{s_1, s_2, ..., s_l\}$
- buggies $B = \{b_1, b_2, ..., b_m\}$
- cars $C = \{c_1, c_2, ..., b_n\}$
- main stations $MT = \{mt_1, mt_2, ...mt_r\}$
- network information $G(V, E)$ where $MT \in V$

while *(Not all passengers are served)* **do**
    foreach *($mt_i \in MT$)* **do**

        - Find first/last mile unassigned passengers $P' \in P$, scooters $S_u \in S$ and buggies $B_v \in B$ such that mt_i is main station for all.
        - Assign $S_u \vee B_v$ to P' s.t. min. cost, f_c.

    end

    - Find middle mile unassigned passengers $P'' \in P$ and cars $c_w \in C$ such that $mt_i == p_k$, where $p_k \in P''$.
    - Assign c_w to $p_k \in P''$ s.t. min. cost, f_c.

end
Greedy: $f_c = t_a(j) + t_t(j), j \in S \vee B \vee C$, where passengers are served based on their request time.
Bipartite: $f_c = \sum_{ij} t_a(i, j) + t_t(i, j) + t_q(i), i \in P, j \in S \vee B \vee C$ where, passengers are served in batches.

Algorithm 1. Multi-class assignment

The input to the algorithm are the road network $G(V, E)$ with edges E represented as the Manhattan distance between the station nodes V, constantly updated location list of unserved or partially served passengers, $P = \{p_1, p_2, ..., p_q\}$, and locations of available multi-class vehicles with the set of scooters as $S = \{s_1, s_2, ..., s_l\}$, buggy set as $B = \{b_1, b_2, ..., b_m\}$ and car set as $C = \{c_1, c_2, ..., b_n\}$. For each iteration of the algorithm, at each main station $mt_i \in MT$, the first and last mile passengers, P' are identified with respect to the current time step. These passengers, P' are assigned either a buggy $b_v \in B_v$ or a scooter $s_u \in S_u$ from the set of buggies and scooters that are available within their respective sub-network. This is followed by assignment of cars $c_w \in C$ for all the middle mile passengers P'' that are at the main stations. When no vehicles are available, the passengers queue up. The assignment is based on the vehicle's accessibility to the network, the assignment time, travel time to destination and passenger's queue time. We implemented two assignment techniques namely, Greedy and Bipartite which are represented by f_c in Algorithm 1. This hierarchical process with first-last mile assignments followed by middle mile assignment is repeated until all the passengers in the system are served. The assignments are performed only when the passengers are either at the main or the sub-station nodes and not along the edges of the network.

As an extension to multi-class assignment with scooters and buggies for first and last miles, we also included walking as the third mode of commute. Ideally, the walking mode needs to be assigned when the sum of queue, travel and assignment time exceeds the walking time as given in Eq. 1.

$$min(t_w, t_q + t_a + t_t) \leq t_w \tag{1}$$

However, for Greedy assignment, this would cause a delay in mode assignment for the passengers as they would initially be required to wait until they are first in the queue and then the assignment time is calculated based on the available vehicles. Hence, we use a heuristic and assign walking mode when the expected queue time, t_q of the passenger exceeds the walking time, t_w.

5   Experimental Setup and Results

We validated our proposed fleet management system in simulation. We considered a real campus environment for testing the three classes of vehicles with different accessibility to the road network and different speeds. Specifically, the main streets on campus are accessible by the self-driving cars, the inter-building paths are navigable by the autonomous buggies and for narrow passageways we used personal mobility scooters. The accessible paths by each category of vehicle were pre-defined. An illustration of our campus map is presented in Fig. 5. The shaded nodes on the map represent the main stations and the plain nodes are sub-stations. The speeds of the vehicles were constantly set to the average speeds of the real platforms with car speed as 5.5 m/s, buggy speed as 3 m/s and scooter speed as 2 m/s. For performance comparison of single-class with

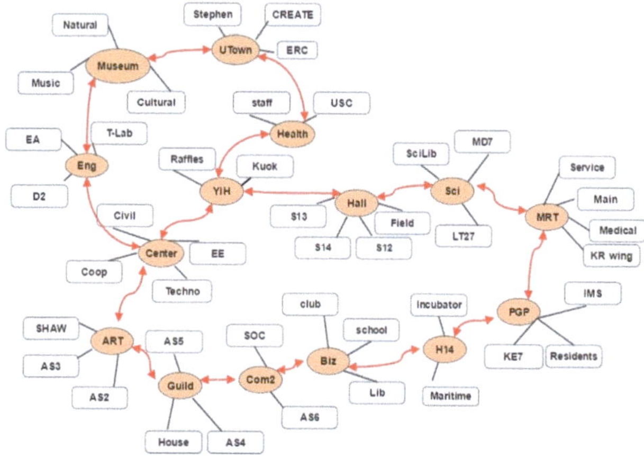

Fig. 5. Case study in National University of Singapore campus.

multi-class assignment combined with walking, we considered the average walking speed as 1.2 m/s. The details of our simulation and field experiments are further explained in the following sections.

5.1   Simulation

5.1.1   Setup

The fleet management system includes simulated passengers, vehicles and booking requests in a real network with the inter-node travel time defined based on the Manhattan distance between the nodes and average speed of the vehicle. Each booking request corresponds to a source and destination location. The source locations are sampled uniformly random from a weighted distribution, where the weights are selected based on the degree of the node in the network. Since, historically the demand is proportional to the connectivity of the node in the campus network. The demand (passenger) inter-arrival time is defined by an exponential distribution with mean based on the average travel time in the network [16]. The destination of the passengers and initial locations of the vehicles are impartially chosen in uniformly random order from the list of stations, while considering the restricted stations for each class of vehicle. The simulation scenario is run until all the passengers are served for their booking requests.

5.1.2   Results

As mentioned earlier, the first challenge for any mobility-on-demand service is to solve the fleet sizing problem. Given the expected demand distribution and the estimated budget, we apply the genetic algorithm to obtain multi-class fleet sizes. The result of applying this algorithm for a budget of approximately $11M$ USD and serving demand size of 1000 passengers in a span of about 40 min is presented in Fig. 6.

46   M. Meghjani et al.

We recorded the best and the mean penalty values for each generation of the genetic algorithm. The penalty value is represented as the total travel time per passenger. The lowest best penalty value corresponding to the lowest total travel time per passenger was evaluated to be 376.982 s using 342 cars, 68 buggies and 302 scooters. The aforementioned quantities obtained correspond to about 91%, 5% and 4% of the total budget for the cost of cars, buggies and scooters, respectively.

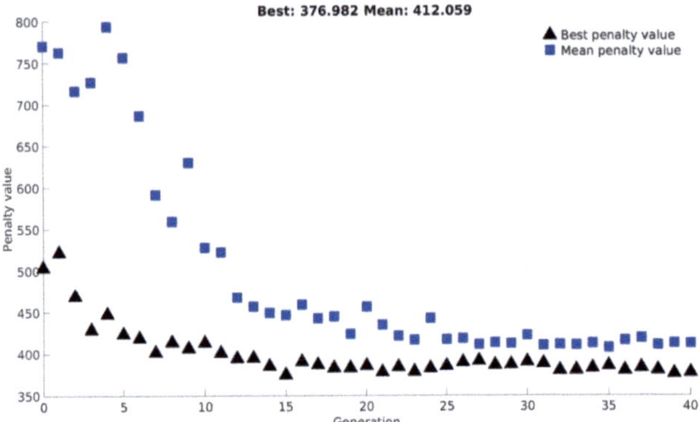

Fig. 6. Total travel time per passenger represented as penalty value corresponding to different combinations of fleet sizes illustrated as generations of the genetic algorithm.

We further analyzed the effect of change in budget (cost) on the total travel time per passenger. For three categories of budget ($6M$, $11M$ and $16M$), we observed that as the travel time decreases, the cost increases exponentially as shown in Fig. 7. This particular relation between the cost and travel time can provide meaningful insights to policy makers to decide on a convenient trade-off between the financial investment for infrastructure and expected travel time of the passengers.

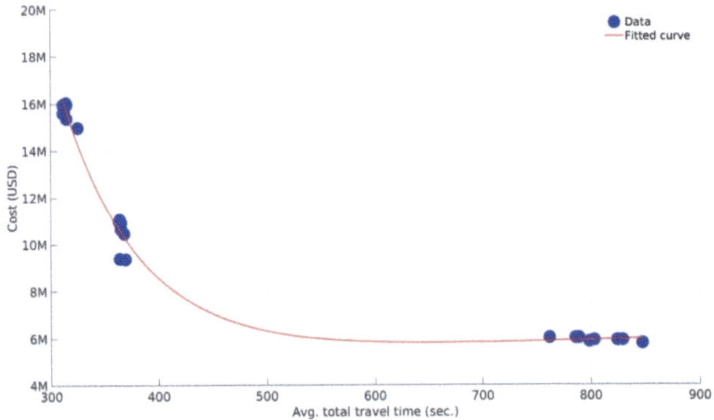

Fig. 7. The total cost of vehicles (budget) varies exponentially with respect to the total travel time per passenger.

Lastly, we compare the total travel time per passenger using our multi-class greedy assignment algorithm to single-class assignment where the passengers walk for the first and last miles. For the single-class assignment, we used the entire budget to purchase all cars resulting in more number of cars for single-class than multi-class. We then divided our analysis into two parts: (a) demand less than 1000 for which we have optimized the fleet size (Fig. 8a) and (b) demand more than 1000 for which the fleet size is insufficient (Fig. 8b). For lesser demand, analysis (a), it is clearly evident that the total travel time per passenger is significantly less for multi-class assignment when compared to single-class. However, when the demand is much greater than the expected demand that we used to optimize the fleet size, it can be observed that the total average travel time for multi-class is worse than the single-class. This is due to the fact that the fleet size is insufficient for the high demand, resulting in longer queue time t_q and hence large total travel time per passenger. In such scenarios, multi-class with walking mode which allows the passengers to walk when the expected queue time is greater than the walking time can significantly reduce the total travel time per passenger.

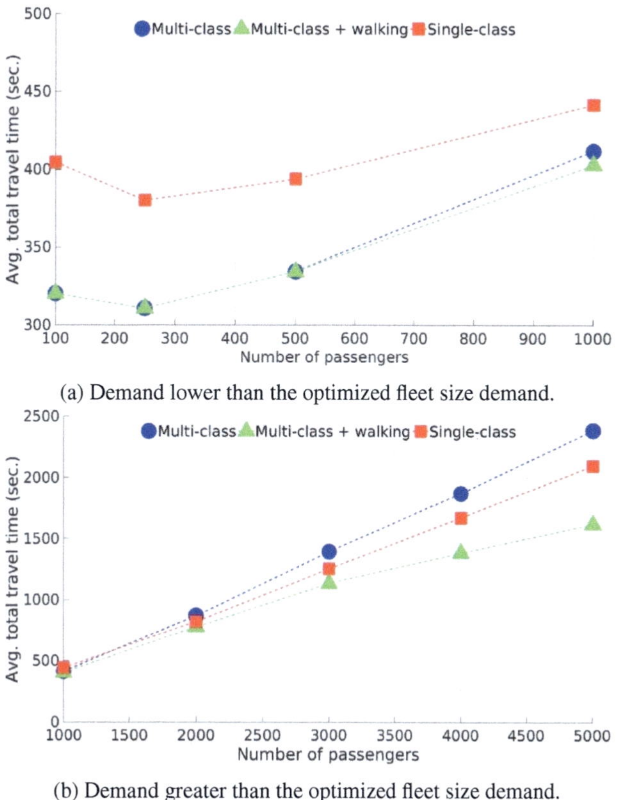

(a) Demand lower than the optimized fleet size demand.

(b) Demand greater than the optimized fleet size demand.

Fig. 8. Multi-class comparison with respect to the total travel time per passenger.

6   Discussion and Conclusion

In summary, we propose a multi-class mobility on-demand system which offers faster transportation than single-class system with the additional convenience of being able to provide door-to-door service and reduced total travel time per passenger. We also included walking mode in our multi-class assignment method which further improved the travel time efficiency of our AMoD system. The travel time efficiency of our multi-class AMoD system depends on the number of vehicles designated in each class. Hence, we used a genetic algorithm to optimize the heterogeneous fleet sizes given an expected budget. Lastly, we empirically analyzed the relation between the budget and total travel time per passenger which can be used for efficient demand management and traffic flow in the network.

Acknowledgment. This research was supported by the National Research Foundation, Prime Minister's Office, Singapore, under its CREATE programme, Singapore-MIT Alliance for Research and Technology (SMART) Future Urban Mobility (FM) IRG. We would also like to acknowledge the support of NVIDIA Corporation's NVAIL program.

References

1. Pendleton, S.D., Andersen, H., Shen, X., Eng, Y.H., Zhang, C., Kong, H.X., Leong, W.K., Ang Jr., M.H., Rus, D.: Multi-class autonomous vehicles for mobility-on-demand service. In: 2016 IEEE/SICE International Symposium on System Integration (SII). IEEE (2016)
2. Zhang, R., Spieser, K., Frazzoli, E., Pavone, M.: Models , Algorithms, and Evaluation for Autonomous Mobility-On-Demand Systems. In: American Control Conference, pp. 2573–2587 (2015)
3. Marczuk, K.A., Soh, H.S., Azevedo, C.M., Lee, D.H., Frazzoli, E.: Simulation framework for rebalancing of autonomous mobility on demand systems. In: International Conference on Transportation and Traffic Engineering (2016)
4. Wang, X.: Optimizing ride matches for the dynamic RDE-sharing systems. Ph.D. thesis, Georgia Institute of Technology (2013)
5. Meghjani, M., Marczuk, K.: A hybrid approach to matching taxis and customers. In: 2016 IEEE Region 10 Conference (TENCON), pp. 167–169, November 2016
6. Azevedo, C.L., Marczuk, K., Raveau, S., Soh, H., Adnan, M., Basak, K., Loganathan, H., Deshmunkh, N., Lee, D.H., Frazzoli, E., Ben-Akiva, M.: Microsimulation of demand and supply of autonomous mobility on-demand. J. Transp. Res. Board (2564), 21–30 (2016)
7. Azevedo, C.L., Marczuk, K., Raveau, S., Soh, H., Adnan, M., Basak, K., Loganathan, H., Deshmunkh, N., Lee, D.-H., Frazzoli, E., Ben-Akiva, M.: Optimizing a flexible mobility on demand system. Transp. Res. Rec. (2536), 76–85 (2015)
8. Psaraftis, H.N.: A dynamic programming solution to the single vehicle many-to-many immediate request dial-a-ride problem. Transp. Sci. **14**(2), 130–154 (1980)
9. Psaraftis, H.N.: An exact algorithm for the single vehicle many-to-many dial-a-ride problem with time windows. Transp. Sci. **17**(3), 351–357 (1983)

10. Cordeau, J.-F., Laporte, G.: A tabu search heuristic for the static multi-vehicle dial-a-ride problem. Transp. Res. Part B Methodol. **37**(6), 579–594 (2003)
11. Google Inc. Google Self-Driving Car Project
12. Montella, C., Perkins, T., Spletzer, J., Sands, M.: To the bookstore! autonomous wheelchair navigation in an urban environment. In: Field and Service Robotics (2014)
13. Pendleton, S., Chong, Z.J., Qin, B., Liu, W., Uthaicharoenpong, T., Shen, X., Fu, G.M.J., Scarnecchia, M., Kim, S.W., Ang, M.H., Frazzoli, E.: Multi-class driverless vehicle cooperation for mobility-on-demand. In: Intelligent Transportation Systems World Congress (ITSWC) (2014)
14. Dantzig, T., et al.: Number, the language of science: a critical survey written for the cultured non-mathematician (1954)
15. Goldberg, D.E.: Genetic Algorithms in Search, Optimization & Machine Learning. Addison-Wesley Publishing Company, Inc. (1989)
16. Allen, A.O.: Probability, Statistics, and Queueing Theory. Academic Press (2014)

SMACOF Hierarchical Clustering to Manage Complex Design Problems with the Design Structure Matrix

Li Qiao$^{(\boxtimes)}$, Mahmoud Efatmaneshnik, and Michael Ryan

Capability Systems Centre, School of Engineering and Information Technology,
The University of New South Wales, Canberra 2600, Australia
{l.qiao,m.efatmaneshnik,m.ryan}@adfa.edu.au

Abstract. Defense system engineering is complex in nature that requires systematic approaches. The design structure matrix (DSM) is a powerful tool for supporting architecture analysis and management of systems. This paper facilitates quantitative analysis by revealing the hidden problem structure. A combined approach using Scaling by MAjorizing a Complicated Function (SMACOF) and hierarchical clustering is proposed to manipulate the design DSM. This algorithm calculates the relevance among the system elements and shows how large problems can be organized into smaller, highly connected topologic modules that combine in a hierarchical manner into larger, less cohesive units. The algorithm also uses Cost and the Jaccard index to guide comparison of results. A simple example is used to illustrate the solution procedure. Also, two real industrial application examples—an aircraft design problem and a satellite multidisciplinary team organization problem—are chosen to demonstrate how the proposed DSM approach manages complexity in the design process.

1 Introduction

The development of systems in defence industry, e.g. aircraft or satellite system design, involves many engineering from different disciplines. For example, the design of a satellite involves a number of subsystem design groups, each of which needs to solve design problems in their particular disciplines. Typical design problems are how to organize various subsystem design groups or their activities to ensure the efficiency of communication and work. To solve this problem, we need to know more about the structure of the design problem since the number of interactions among various sub-problems can lead to a significant level of complexity. Therefore, we need systematic approaches to understand the problem and managing the complexity. Knowledge about the structure of the design problem is important for deepening our understanding of design. If we can find a way to trace the structure of design problems, and we can match that to the way designers tackle those problems, this will open the possibility for a much closer description, and a better understanding of the way designers work and why take the actions we see [1].

© Springer Nature Switzerland AG 2019
M. A. Cardin et al. (Eds.): CSD&M 2018, AISC 878, pp. 50–61, 2019.
https://doi.org/10.1007/978-3-030-02886-2_5

The Design Structure Matrix (DMS) is a matrix representation of a complex product or a system [2]. DSM has proved to be an efficient matrix-based tool to support architecture analysis, and to perform both analysis and management of complex problems [3]. Since the DSM is a matrix representation of a complex system, users can model, visualize and analyze the dependencies among the system elements and derive suggestions for the improvement or synthesis of the integrated system. There are three main types of DSM: component-based DSMs that represent component relationships are used for product architecting; people-based DSMs that represent organizational relationships are used for organization design and interface management; activity-based DSMs which represent process relationships are used for project scheduling management. DSM focuses on dependencies that correspond to the entries in the matrix. Depending on the rating schemes employed, the entries in DSM can be logical/binary or real numbers to reflect the strength of the interactions between the system elements.

Common methods to manipulate DSMs include partitioning, clustering and tearing. Reference [3] introduces various DSM methods and applications. To our best knowledge, only two papers have discussed the application of multidimensional scaling (MDS) clustering to DSM. In the first paper [4], SMACOF (Scaling by MAjorizing a Complicated Function)/Fuzzy C-Means is a combination under MDS clustering strategy that successfully identifies the platform for a hydraulic support system. In order to verify the effectiveness of MDS clustering strategy, the second paper, i.e. our previous work [5], provides a rigorous study of its application to DSM modular analysis. Optimal modularity exists for systems [6] and our work in has shown that MDS clustering is capable of finding such optimal modularization. We reveal two advantages of MDS clustering methods. First, these methods outperform the Newman-Given algorithm, which has been extensively applied to DSM clustering. Second, these methods are capable to produce the optimal modules for any number of clusters, which is favourable especially when the cluster number is a higher managerial decision. However, our previous work focused on the evaluation of the performance of MDS clustering to reveal the most suitable combination setup for DSM application. That study only targets the numbered DSM, but does not study the logical DSM.

In this paper, we consider tailoring of our previous algorithm for the structuring of design problems. To be specific, we use the combined SMACOF and hierarchical clustering using *Cosine* as a distance metric to manipulate the DSM, marked as DSM SMACOF hierarchical/*Cosine*. The procedure has four-stages: (1) build DSM that represent design problems, (2) use SMACOF to produce the embedded data, (3) hierarchically cluster the data, (4) evaluate and deploy knowledge. The proposed method is evaluated using two existing examples (respectively represented by logical and integer DSM) from literature as study cases.

The remaining of this paper is structured as follows. Section 2 presents the problem formulation derived from the relevant work; Sect. 3 presents the detailed description of the procedure using a simple example for ease of understanding; Sect. 4 evaluates this method using two real industrial design problems; and Sect. 5 draws some conclusions.

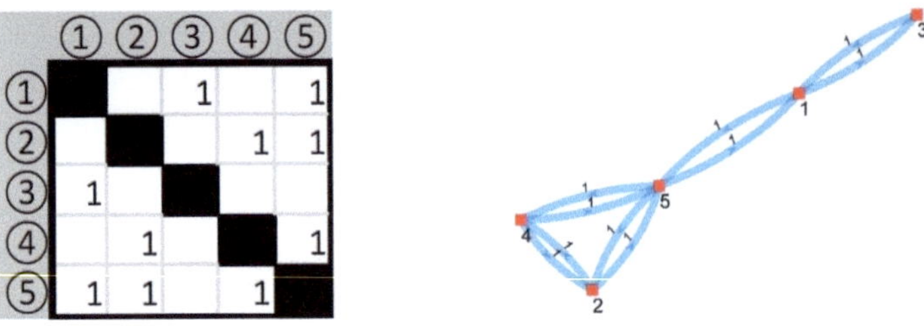

Fig. 1. A binary DSM5 (left) and its graph (right)

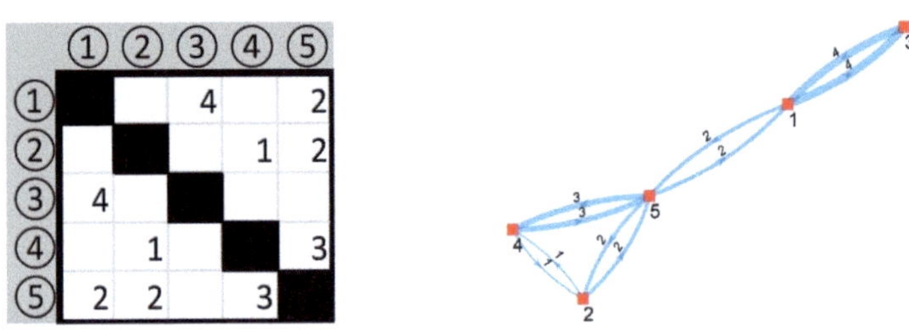

Fig. 2. A integer DSM5 (left) and its graph (right)

2   Problem Formulation

As explained earlier, we extend our previous work in this paper. Under the MDS clustering strategy, we find the effective setup here is the combination of the SMACOF and hierarchical/*Cosine*. Since the focus is on the application, we only provide a brief explanation to the DSM, SMACOF, hierarchical clustering and the formulas for clustering evaluation.

2.1   DSM and Its Graph Representation of a Design Problem

A complex design problem can be represented by a graph of interactions. A directed graph is a set of n nodes and associated edges. In many applications, each edge of a graph has an associated numerical value (which is usually taken to be a positive integer and we only consider positive integers in this study). According to the graph theory, the relationships between elements can be mapped to a matrix, called a DSM [3]. If there is a directed flow from element i to j in the graph, there is a positive integer in the $(i, j)th$ cell (row i, column j), and a 0 in the matrix cell otherwise. Any graph has its corresponding DSM.

Hereafter, we use the notation DSMn for a DSM with n elements. Assume a problem consists of 5 sub-problems, which can be people, activities, processes, etc. Then we represent the problem with a 5-element graph, using Node $1, 2, \cdots 5$ to number the sub-problems. Figure 1 shows a binary DSM and its logical graph, where designers place binary numbers (1s and 0s) to indicate the presence and absence of the relations. Figure 2 shows an integer DSM and its weighted graph, where designers assign strength ranking numbers (e.g. 1, 2, 3 ...) on the edges to present the strengths of interactions such as the effect or magnitude of dependencies. It can be seen that we can visualize the interactions between the subproblems using a DSM and its graph: a binary DSM on a logical level, and an integer DSM on a detailed level.

2.2 SMACOF (One Type of MDS Techniques)

MDS refers to a class of techniques that use proximities of objects as input, also called low-dimensional embedding or an ordination method [7]. The output of MDS - the embeded data - represent a set of objects in a few dimensions while preserving their multidimensional inter-item distances/similarity as close as possible. SMACOF is one type of MDS techniques, which provides an approach for multidimensional scaling based on stress minimization by means of majorization. Details of SMACOF can be found in [8]. DSM of a complex system is a large n-dimensional data set. We use SMACOF to preprocess the original data before clustering. The output of SMACOF is a m-dimensional set of embedded data. The most important benefit of this data pre-processing is that the computational load decreases considerably, making it possible to cluster large data sets in a limited time. The clusters found using the embedded data are similar to those of the original data [9].

2.3 Agglomerative Hierarchical Clustering

Cluster analysis is a type of unsupervised machine learning that enables industries to discover the hidden structure of a complex system. The discussion of different clustering approaches is presented in [10]. Here, we investigate the use of the agglomerative hierarchical clustering. Agglomerative hierarchical clustering builds a hierarchy structure of clusters by merging all objects, producing a single, all-inclusive cluster at the top, and singleton clusters of individual objects at the bottom. First, each of the n objects to be clustered is considered as a unique cluster. Then, the objects are compared among themselves using a measure of distance. *Cosine*, the angular difference of two data vectors (each data vector presents a system element), is used as a distance metric in the present study. Next, the two clusters with the smaller distance are joined. This process iterates until all objects form a large single group that consists of all objects. A linkage method is used to compare the clusters at each stage and to decide which of them should be combined. Common procedures are single, complete and average linkage. Here the widely used average linkage is selected. For more detail about hierarchical clustering refer to [10].

Table 1. Algorithm

Algorithm	SMCOF hierarchical clustering with DSM
Input:	DSM (an $n \times n$ matrix)
Output:	Dendrogram, Partitions for a range of k with Cost, optimal partitions
Step 1: DSM and its digitalization	**If** the entries are not numbers, **then** digitize the DSM, **end if.**
Step 2: SMACOF	Use SMACOF, obtain embedded data (an $n \times m$ matrix)
Step 3: Hierarchical clustering	Compute the distance matrix using *Cosine*, hierarchical clustering the embedded data, obtain the partitions for a range of k
Step 4: Evaluation and knowledge deployment	Calculate the *Cost* for a range of k **If** the P_{ref} exits = True Calculate the *Jaccard* index between the P_{ref} and the obtained partitions **end** **If** there are no design constraints **then** choose the optimal partition when $k = k^*$ **elseif** there are design constraints, modify the partition **end if**

2.4 Evaluation and Knowledge Deployment

We continue to use the *Cost* function in Eq. 1 to compare the quality of the partition solutions as we did in [5].

$$
\begin{aligned}
Cost &= \sum IntraClusterCost + \sum ExtraClusterCost \\
&= \underbrace{\sum [DSM(i,j) + DSM(j,i)] \times d_k}_{i,j \text{ are in the same cluster}} + \underbrace{\sum [DSM(i,j) + DSM(j,i)] \times n}_{i,j \text{ are not in the same cluster}} \quad (1)
\end{aligned}
$$

The left summation accounts for the interactions that are within the clusters, by adding the number of elements of the corresponding cluster C_k, denoted by d_k. The right summation considers the interactions outside the clusters by adding the size of the full matrix (i.e. n) to the cost of each interaction. When we calculate the *Cost* for a range of partition solutions, the *Cost* value first decrease as the number of cluster increase, which is caused by the decrease in the cluster's sizes. Then, the *Cost* value increase, as a result of an increase in the number of interactions outside the partitions. Therefore, the minimum value for the *Cost* is considered as the optimal partition. Good clustering results in inappropriate partitions with low cost. Based on that, we can find the preferred solution as the optimal partition with the lowest *Cost* value. The goal here is not only to find an optimal clustering for the data, but also to obtain good insights in the cluster structure of the data. Thus, as the extension to our previous work, we try to gain more knowledge through two kinds of comparison: (1) comparison of the obtained partitions with the reference partition P_{ref} and (2) comparison within the obtained partitions. The *Jaccard* index in Eq. 2 is used to compare the partitions. X and Y are two sets of the members, $|X \cap Y|$ is the number of members which are shared between both sets, and $|X \cap Y|$ is the total number of members in both sets (shared and un-shared). The *Jaccard* index of two identical partitions is 1.

$$
Jaccard(X,Y) = |X \cap Y|/|X \cap Y| \quad (2)
$$

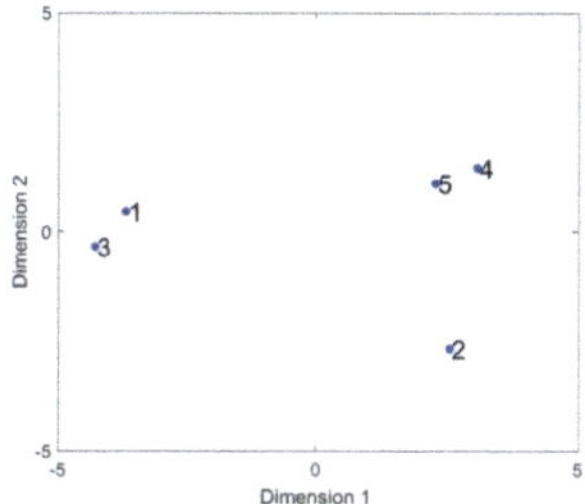

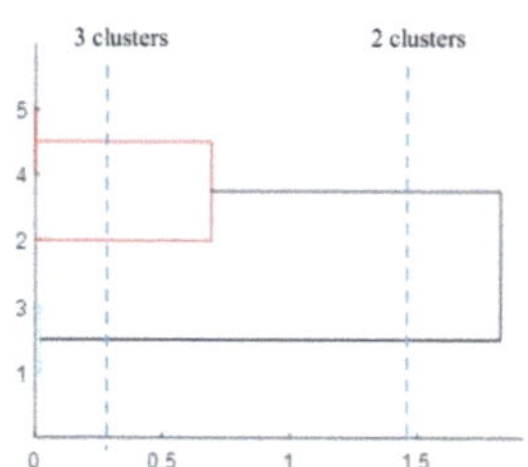

Fig. 3. Representation in new 2D by MDS

Fig. 4. Dendrogram representing nested clusters for DSM5. A partition can be obtained by cutting the dendrogram at a certain level.

3   Procedure

For the ease of understanding, the simple DSM5 $(n = 5)$ in Fig. 2 is used to present the algorithm shown in the Table 1. When the entries of a DSM are not expressed in numbers, e.g. using symbol or size of dots, the original DSM should be digitized to a binary or an integer DSM.

According to [5], we set the value of $m = fix(n/2)$, rounding $n/2$ to the nearest integer toward zero. Here, $m = 2$, thus we could show the embedded data in a 2D space (see Fig. 3). A number of aspects are intuitively obvious from the distribution. For instance, Node 1 and 3 are close to each other which agrees with our observation in Fig. 2. Note that most systems which consist of a large number of elements (n is not a small number), will not have such low dimensional representation. The original data is a 5×5 matrix, and the embedded data is a 5×2 matrix. This new matrix is the input for the following hierarchical clustering. Note that the visualization in Figs. 2 and 3 can only be used to obtain some qualitative information. In contrast, we obtain a quantitative description of data properties, such as the measure of the closeness of Node 1 and 3. The solutions of hierarchical clustering (see Fig. 4) are in the form of trees, called dendrograms. The horizontal axis represents the dissimilarity between elements. The vertical axis represents the elements. We can see that Node 1 and 3 are more similar

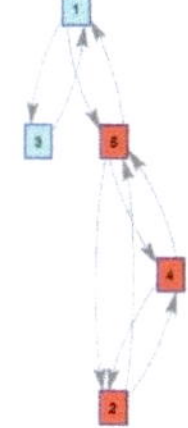

Fig. 5. A integer DSM5 (left) and its graph (right)

to each other than they are to Node 4, 5 and 2. A partitioning can be achieved by cutting the dendrogram at certain level(s), such as $P_{k=2} = \{\{1,3\}, \{2,4,5\}\}$ and $P_{k=3} = \{\{1,3\}, \{4,5\}, \{2\}\}$ (see Fig. 4). Each *Cost* corresponds to a specific partitioning. The *Cost* value of $P_{k=2}$ and $P_{k=3}$ is 72 and 78, respectively. We found that the optimal $k^* = 2$. If there are no design constrains, we can suggest the solution preference is $P_{k^*=2}$, and its corresponding segmented DSM and graph are shown in Fig. 5. This is more useful for designers compared to Fig. 2.

4   Industrial Case Study

In this section, we use two case studies to demonstrate how the proposed approach support analysis the complex problems. The first case from [11] is a team-based (people-based) satellite design problem, which DSM is a binary matrix. The second case from [12] is an aircraft design problem, which DSM is an integer matrix. Our focus is on how to interpret and gain new knowledge from the results.

4.1   Team-Based Satellite Design Problem

Figure 6 shows a team-based DSM of 16 subsystems/disciplines in a satellite design problem, marked as DSM16. The mark "X" presents the interaction among members of the design team. We represent the problem with a binary DSM by replacing "X" with 1. It is hard to find any structure characteristics by direct observations. In [11], the New-Girvan algorithm is adopted for clustering, and the result has six clusters, as shown in Fig. 6.

We first build a cluster hierarchy (see Fig. 7). This dendrogram helps to identify the related elements, such as Thermal (12) and Electrical power (14). Then, we plot a diagram of *Cost* versus k (see the dotted line in Fig. 8) and the *cost* of P_{ref} (see the red solid line in Fig. 8). Compared to P_{ref}, our result has lower cost of 264. We find the optimal clustering for the optimal number of cluster $k^* = 4$. We use the bordered squares to represent clusters of the matrix and scale the size of the dots according to the value of entities.

We compare the obtained 7 partitions $(P_{k=2}, \ldots, P_{k=8})$ to the P_{ref} with *Jaccard* index, shown in Fig. 10. The partition $P_{k=5}$ is the most similar one to the P_{ref}. Comparing $P_{k=4}$ to $P_{k=5}$, The difference is that Thermal (12) is separated from the cluster that consists of Avionics (5), Flight software (6), Reliability (7), Electrical power (14). The two partitions are as follows:

$$P_{k=4} = \{\{2,3,4\}, \{5,6,7,12,14\}, \{8,9,10,11,13\}, \{1,15,16\}\}$$
$$P_{k=5} = \{\{2,3,4\}, \{5,6,7,14\}, \{12\}, \{8,9,10,11,13\}, \{1,15,16\}\}$$

Launch Vehicle (1), Integration and Test (15) and Parametric Cost (16) are included in this DSM though there is no dependency in the rows and columns corresponding to those disciplines. Let us set the design constraints here that (1), (15) and (16) must be located in independent groups. Then we can revise

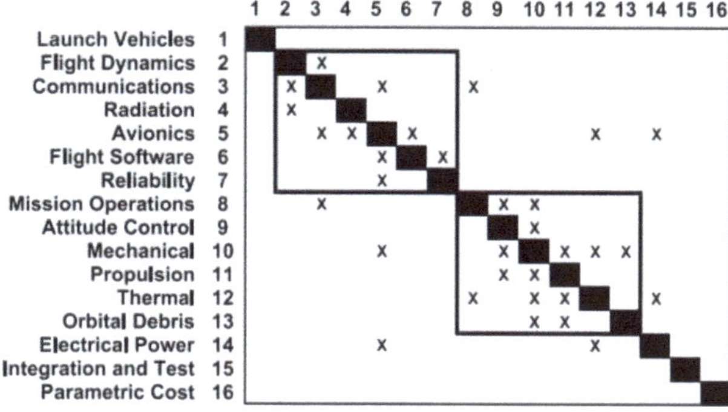

Fig. 6. Segmented DSM16 for the satellite design process [11]. The P_{ref} is $\{\{1\},\{2,3,4,5,6,7\},\{8,9,10,11,12,13\},\{14\},\{15\},\{16\}\}$

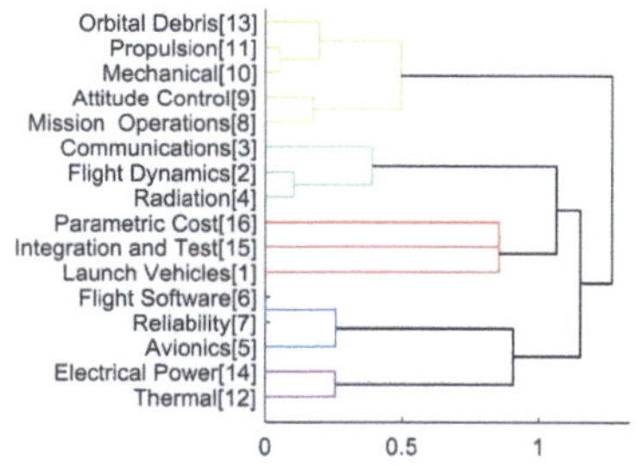

Fig. 7. Dendrogram representing nested clusters for DSM16

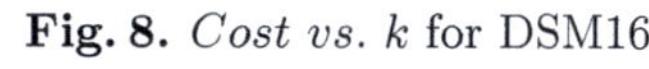

Fig. 8. *Cost vs. k* for DSM16

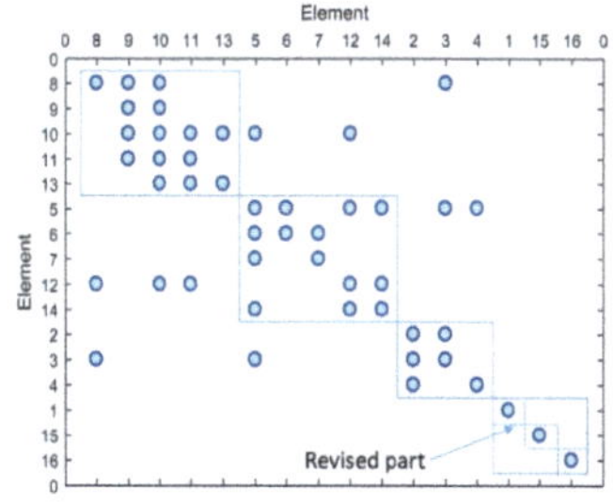

Fig. 9. Dendrogram representing nested clusters for DSM16

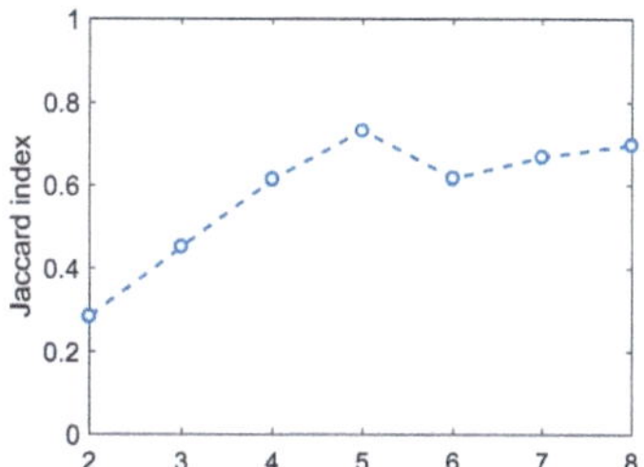

Fig. 10. *Cost vs. k* for DSM16

the partition with the same lowest *Cost* value (see the dotted line in Fig. 9). We use a colour-coded graph to indicate the hierarchical modularity of the optimal partition (see Fig. 11). The teams with the same colour should work more closely together.

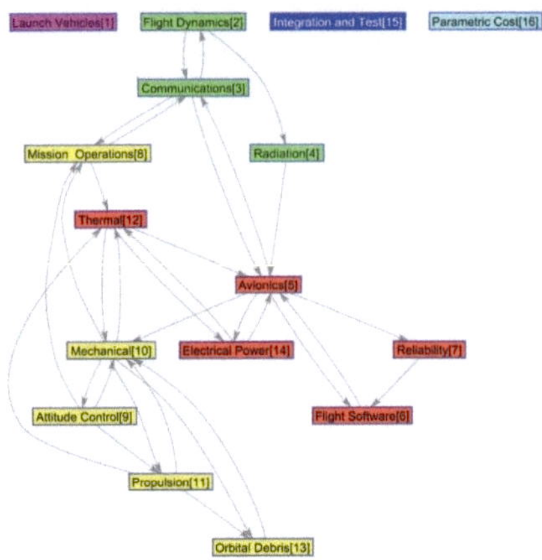

Fig. 11. Colour-coded graph of the revised segmented DSM16

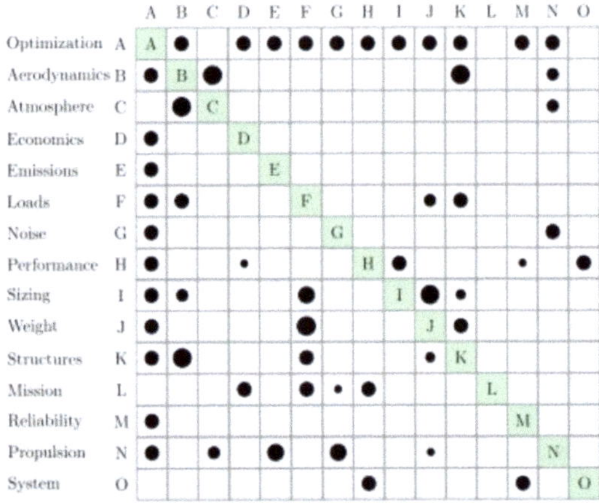

Fig. 12. Example DSM for aircraft design problem [12]

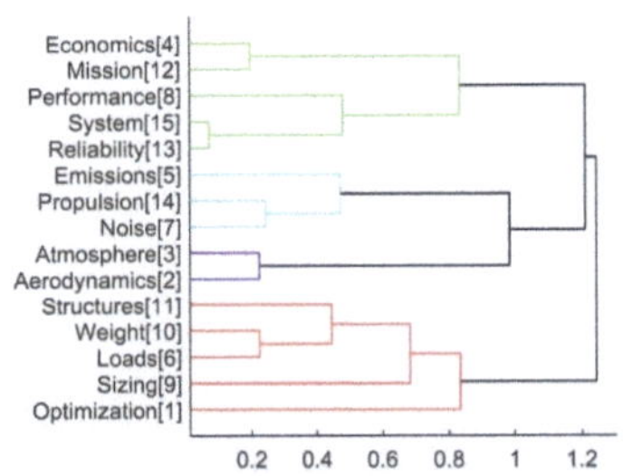

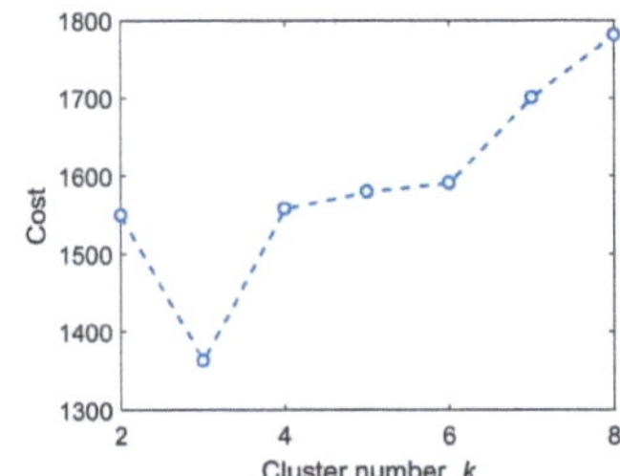

Fig. 13. Dendrogram representing nested clusters for DSM15

Fig. 14. *Cost vs. k* for DSM15

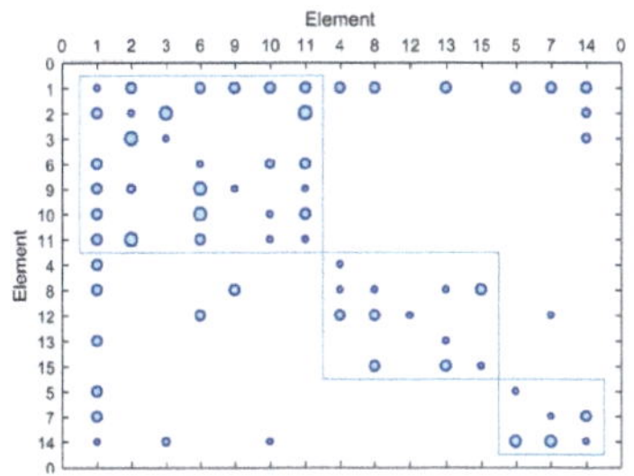

Fig. 15. Segmented DSM15 with $P_{k=3}$

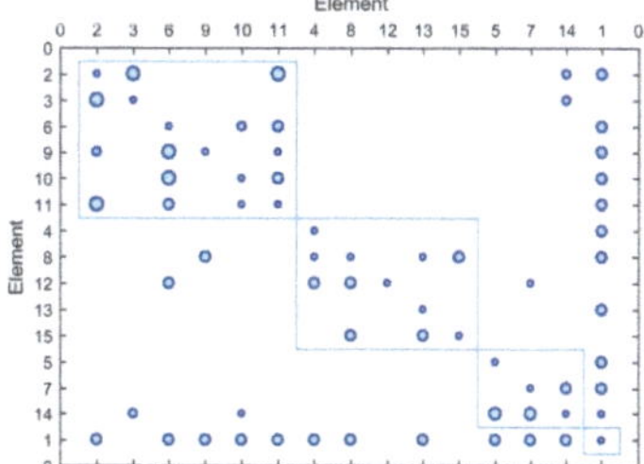

Fig. 16. Segmented DSM15 with $P_{k=4}$

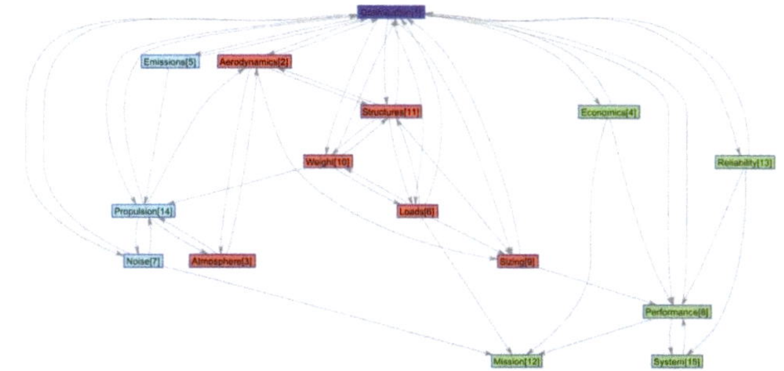

Fig. 17. A colour-coded graph represented for the segmented DSM15

4.2 Activity-Based DSM for Aircraft Design Problem

Reference [12] is an example of an aircraft design problem, shown as DSM15 in Fig. 12. There are 15 elements, presenting various decision-making activities of the whole design problem. The entities represent the information flow between elements, where larger dots denote stronger coupling between the disciplines. Here, we assign an integer ranging from 1 to 5 to replace the dots according to the size of dots.

Figure 13 shows the dendrogram representing nested clusters for DSM15. For instance, Weight (10) and Loads (6) are tightly connected, then they merged with Structures (11). Atmosphere (3) and Aerodynamics (2) are highly coupled. Figure 14 shows the plot of $Cost vs. k$, which suggests the optimal is $k^* = 3$. Figure 15 presents the segmented DSM15 with $P_{k=3}$. However, it is not satisfied as Optimization (1) is a bus-like element which has links to most of the rest activities. Thus we look for a partition with low cost and in which the Optimization (1) is isolated. Then we revised our preferred partition to $P_{k=4}$ (see Fig. 16). $P_{k=4}$ is more reasonable as if we remove Optimization (1), the rest elements group very well as very few interactions are outside the cluster. Figure 17 is the corresponding colour-coded network of $P_{k=4}$.

5 Conclusion

Knowledge about the structure of a complex design problem is important for deepening our understanding of design. Our aim was to facilitate quantitative analysis to obtain the insights of the design problem that remain hidden to the analyst, by uncovering hidden patterns and structures within the problem.

We tailored our previous study for this specific application. A DSM-based SMACOF hierarchical clustering method is proposed to model, analyze, and manage the complexity of system design problems. One simple example is used to show the detailed procedure of our proposed algorithm. Two real industrial study cases (respectively represented by logical and integer DSM) are applied. The proposed algorithm is found to perform well and capable to build a cluster hierarchy (dendrogram) for a large problem that combines various sub-problems; and to illustrate the patterns of problem-related behaviour emerge, without being completely problem-specific. Using the *Cost* value, the algorithm can determine the optimal modularity as a preference solution. The use of the optimal clustering can enhance efficiency of the problem solving, and prevent both project management and technical failures. In addition to *Cost*, the *Jaccard* index is used to compare partitions and reveal more insights on different partition solutions. The study has demonstrated that the proposed algorithm can successfully support the knowledge deployment process for the complex design problems.

References

1. Dorst, K.: The problem of design problems. In: Expertise in Design, pp. 135–147 (2003)
2. Browning, T.R.: Applying the design structure matrix to system decomposition and integration problems: a review and new directions. In: IEEE Trans. Eng. Manage. **48**(3), 292–306 (2001)
3. Eppinger, S.D., Browning, T.R.: Design Structure Matrix Methods and Applications. MIT press, Cambridge (2012)
4. Li, Z., Cheng, Z., Feng, Y., Yang, J.: An integrated method for flexible platform modular architecture design. J. Eng. Design **24**(1), 25–44 (2013)
5. Qiao, L., Efatmaneshnik, M., Ryan, M., Shoval, S.: Product modular analysis with design structure matrix using a hybrid approach based on MDS and clustering. J. Eng. Design **28**(6), 433–456 (2017)
6. Efatmaneshnik, M., Ryan, M.J.: On optimal modularity for system construction. Complexity, **21**(5), 176–189 (2016)
7. Hofmann, T., Buhmann, J.: Multidimensional scaling and data clustering. In: Advances in Neural Information Processing Systems, pp. 459–466 (1995)
8. Borg, I., Groenen, P.J.F.: Modern Multidimensional Scaling: Theory and Applications, 2nd edn. Springer Science & Business Media, New York (2005)
9. Vesanto, J., Alhoniemi, E.: Clustering of the self-organizing map. IEEE Trans. Neural Netw. **11**(3), 586–600 (2000)
10. Aggarwal, C.C., Reddy,C.K.: Data Clustering: Algorithms and Applications. CRC press (2013)

11. Avnet, M.S., Weigel, A.L.: An application of the design structure matrix to integrated concurrent engineering. Acta Astronautica **66**(5-6), 937–949 (2010)
12. Lambe, A.B., Martins, J.R.R.A.: Extensions to the design structure matrix for the description of multidisciplinary design, analysis, and optimization processes. Struct. Multidisciplinary Optim. **46**(2), 273–284 (2012)

Modelling Safe and Secure Cooperative Intelligent Transport Systems

Giedre Sabaliauskaite[(✉)], Jin Cui, Lin Shen Liew, and Fengjun Zhou

Centre for Research in Cyber Security (ITrust), Singapore University of Technology and Design, Singapore 487372, Singapore
{giedre,jin_cui,linshen_liew,fengjun_zhou}@sutd.edu.sg

Abstract. Automated Vehicles (AVs) are expected to help in significantly reducing traffic injuries and fatalities in the near future. However, to achieve this goal, they must be safe and secure. The recent news of the first fatal crash of AV including pedestrian confirm the urgent need of addressing AV safety and security issues to prevent such accidents from happening in the future. In order to outperform human drivers, AVs need to communicate with the other traffic participants, which makes them more vulnerable to cyberattacks. Cooperative Intelligent Transport Systems (C-ITS), which include vehicle-to-vehicle and vehicle-to-infrastructure communications, are expected to be launched in Europe next year. Thus, assuring their safety and security is crucial. This paper presents an approach, CESAM&SSM, for modelling safe and secure C-ITS using the CESAM method and the Six-Step Model. A combination of these two methods enables comprehensive analysis of C-ITS from operational, functional, constructional, safety, and security perspectives. The propose approach is compliant with three international standards: ISO 26262 – vehicle safety standard, SAE J3061 – vehicle cybersecurity standard, and ISO 21217 – intelligent transport system architecture standard.

Keywords: Cooperative intelligent transport system · Automated vehicle
Safety · Security · CESAM method · Six-Step model

1 Introduction

Nowadays, Automated Vehicles (AVs) - the self-driving vehicles - are becoming a reality. In AVs, the automated driving system is able to partially or even completely replace a human driver in performing the driving functions required to operate the vehicle in on-road traffic. AVs can help to reduce commuting time, enable more people to enjoy freedom of traveling (e.g. elderly and people with disabilities), and reduce traffic injuries and fatalities (NHTSA 2017).

However, to fully achieve these goals, AVs must be safe and secure. Unfortunately, the first fatal crash of an AV including pedestrian has been reported in March 2018 (The Guardian 2018). This will undoubtedly increase the emphasis on the urgent need to assure AV safety and security to prevent such accidents from happening in the future.

© Springer Nature Switzerland AG 2019
M. A. Cardin et al. (Eds.): CSD&M 2018, AISC 878, pp. 62–72, 2019.
https://doi.org/10.1007/978-3-030-02886-2_6

AVs are complex Cyber-Physical Systems (CPSs), which integrate embedded computing technology into physical phenomena, and therefore they are vulnerable not only to failures, but also to cyber-attacks. Thus, safety and security have to be considered while developing, testing, and deploying AVs on public roads (NHTSA 2017).

In order to outperform human drivers, AVs need to communicate with the other traffic participants. AVs without connectivity neither can "see" the vehicles several positions further ahead nor can anticipate the actions of the other vehicles. The communications will allow road users and traffic managers to share and use information previously not available and to coordinate their actions (EC 2016). The systems of connected cooperative AVs, which include AVs, roadside infrastructure, and other systems, are called Cooperative Intelligent Transport Systems (C-ITS) (EC 2016, Sjoberg et al. 2017).

In Europe, a CAR-2-CAR Communication Consortium (C2C-CC) has been established with the primary objective of further increasing road traffic safety and efficiency by means of C-ITS (Car2Car 2015). C2C-CC has defined a 4-phase roadmap for deployment of C-ITS: awareness driving phase (vehicles disseminate only their status information), sensing driving phase (vehicles exchange their sensor information), cooperative driving phase (vehicles share their intentions with other traffic participants), and, finally, synchronized cooperative driving phase (vehicles exchange and synchronize their driving trajectories to achieve optimal driving patterns) (Sjoberg et al. 2017).

C-ITS are vulnerable to failures and attacks just as AVs. Thus, assuring safety and security of C-ITS is crucial. How can we analyze C-ITS safety and security in a consistent way throughout the entire development life-cycle, taking into consideration above-mentioned C-ITS deployment phases?

In our previous work, we proposed an approach for AV safety and security analysis, which uses a Six-Step Model for integrating and maintaining consistency among safety and security processes and artefacts of an AV at a single-vehicle level (Sabaliauskaite and Cui 2017). This approach is compliant with the international vehicle safety and cybersecurity standards, namely ISO 26262 "Road vehicles – functional safety" (ISO 26262-3 2011) and SAE J3061 "Cybersecurity guidebook for cyber-physical vehicle systems" (SAE J3061 2016a, b).

In this paper, we extended the initial approach to the AV system-of-systems level to enable modelling of safe and secure C-ITS. The proposed approach, CESAM&SSM, integrates the Six-Step Model (SSM) with the complex system architecting method CESAM (CESAM 2017). In addition to the international vehicle safety and cybersecurity standards ISO 26262 and SAE J3061, it takes into consideration the C-ITS deployment phases and the international standard ISO 21217 (Intelligent transport systems – communication access for land mobiles (CALM) - architecture) (ISO 21217 2014).

The remainder of the paper is structured as follows. Section 2 describes preliminaries. Section 3 explains AV safety and security. Section 4 describes the proposed approach. Finally, Sect. 5 concludes the paper.

2   Preliminaries

2.1   Automated Vehicle Main Terms and Definitions

Automated road vehicles perform the driving functions required to operate the vehicle in on-road traffic. These are the real-time operational and tactical functions, which include lateral and longitudinal vehicle motion control, monitoring the driving environment, object and event response execution, maneuver planning, and enhancing conspicuity via lighting, signaling, etc. These functions are collectively called the Dynamic Driving Task (DDT) (SAE J3016 2016a, b).

AVs perform entire or part of DDT depending on their automation level. International standard SAE J3016 (2016a, b) describes six driving automation levels. At level 0, the human driver performs entire DDT. At level 1, an automated system can assist the human driver to perform either the lateral or the longitudinal vehicle motion. At level 2, an automated system performs the lateral and the longitudinal vehicle motion, while driver monitors the driving environment. At level 3, an automated system can perform entire DDT, but the human driver must be ready to take back control when the automated system requests. There is no human driver at level 4; an automated system conducts the entire DDT, but it can operate only in certain environments and under certain conditions. Finally, at level 5, an automated system performs entire DDT in all environments.

AVs implement DDT using a set of functions, which can be grouped into three main categories: perception (perception of the external environment/context in which vehicle operates), decision & control (decisions and control of vehicle motion, with respect to the external environment/context that is perceived), and vehicle platform manipulation (sensing, control and actuation of the vehicle, with the intention of achieving desired motion) (Behere and Törngren 2016).

AVs consist of two main systems: cognitive driving intelligence system (implements perception and decision & control functions) and vehicle platform (implements vehicle platform manipulation function). Each system can be further decomposed into subsystems and components, which include hardware, software, and communications (Behere and Törngren 2016).

2.2   The Six-Step Model for Complex System Modeling

Complex system modeling has been researched since as early as 1960s. Most complex systems are formed as hierarchies, such as functional, structural, behavioral, etc. (Modarres and Cheon 1999). Early research in this area focused on complex physical systems. GTST-MLD, a function-centered approach as a framework for modeling complex physical systems, has been proposed by Modarres and Cheon (1999). It comprises of Goal Tree-Success Tree (GTST) and Master Logic Diagram (MLD). GTST is a functional hierarchy of a system organized into different levels. The role of GT is to describe system functions, while ST is aimed at describing the structure (configuration) of the system, used to achieve functions identified in GT. Finally, MLD is used to model the interrelationships between functions (GT) and structure (ST).

Brissaud et al. (2009) extended GTST-MLD by integrating faults and failures into it. A new framework was named the 3-Step Model. It allowed modeling the relationships between faults and failures, and the system functions and structure. The analysis of these relationships could be used to assess the effect of any fault or failure on any material element and/or function of the system.

GTST-MLD and the 3-Step Model are insufficient for modelling CPSs, which, in addition to faults and failures, are exposed to cybersecurity threats. Thus, in our previous research we extended the 3-Step Model and proposed a Six-Step Model for integrated CPS safety and security modeling and analysis (Sabaliauskaite et al. 2017). It incorporates cyber-attacks together with safety and security countermeasures into the 3-Step Model, and enables the modeling of interrelationships between failures, attacks, safety and security countermeasures, and system functions and structure. Its construction consists of the following six steps (see Fig. 1):

1. *Functional hierarchy.* The functions are defined using the Goal Tree (GT), which is constructed starting with the goal (functional objective) and then defining functions and sub-functions, needed for achieving this goal. A relationship matrix, F-F, is used to define the relationships between functions, which can be high, medium, low, or very low.
2. *Structural hierarchy.* In the second step, a Success Tree (ST) is used to describe system's structure as a collection of sub-systems and units. Furthermore, the relationships between structure and functions are defined using a relationship matrix S-F.
3. *Failure hierarchy.* The third step is focused on safety hazard analysis, where system's failures are identified and added to the model. In addition, the relationships between failures, system structure and functions are identified, and the corresponding relationship matrices – B-B, B-S, and B-F are constructed.
4. *Attack hierarchy.* The fourth step focuses on security threat analysis, where attacks are identified and added to the model along with the relationship matrices to describe relationships between attacks, failures, structure and functions. Relationship matrix A-B is used to determine which failures could be triggered by a successful attack.
5. *Safety countermeasures.* In this step, safety countermeasures are added to the model and their relationships are identified. Matrices X-A and X-B show the coverage of attacks and failures by safety countermeasures, where white rhombus indicates that the countermeasure provides low protection from attack/failure; gray rhombus - medium protection; black rhombus - full protection.
6. *Security countermeasures.* In the final step, security countermeasures are added to the model and their relationships are established. Matrices Z-A and Z-B are added to define the coverage of attacks and failures by security countermeasures. The security countermeasures, added in this step, could be used to protect the system from attacks and failures, not covered by the safety countermeasures. Furthermore, matrix Z-X is used to capture the inter-dependencies between safety and security countermeasures (reinforcement, antagonism, conditional dependency, and independence), as defined by Piètre-Cambacédès and Bouissou (2010).

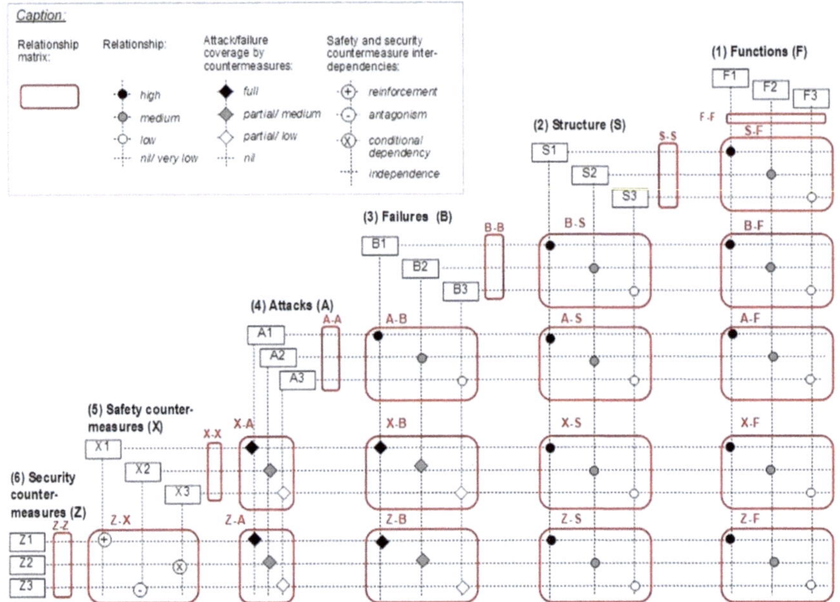

Fig. 1. The Six-Step model.

After completion of each step, it is important to analyze if there were any changes made to previous steps. If the changes occurred, it is necessary to repeat the previous steps.

2.3   Connected Vehicles and Cooperative ITS (C-ITS)

Cooperation and coordination among AVs and other traffic participants is becoming increasingly important with the development of highly automated vehicles (EC 2016, Sjoberg et al. 2017). AVs continually plan and choose their trajectories based on the observed environment. An AV without connectivity can only see the vehicles immediately adjacent to them. Since the plans of the other traffic participants are not known, AVs must include large buffers in their trajectories to deal with the uncertainties. The communication among AVs would enable them to drive closer to each other, operate with better control and have quicker reaction, and eventually avoid collisions.

C-ITS enable AV cooperation and coordination using vehicle-to-vehicle (V2V) and vehicle-to-infrastructure (V2I) communications, collectively called V2X (EC 2016, Sjoberg et al. 2017) and provide various services, such as hazard and vulnerable road user warnings, lane-merging assistance, platooning, etc. C-ITS are expected to be launched in Europe in 2019 (EC 2016, Sjoberg et al. 2017) and deployment in four phases (see Fig. 2):

1. Awareness Driving phase (vehicles disseminate their status information allowing other vehicle to be aware of the presence of other vehicles and hazards);

2. Sensing Driving phase (in addition to status data, vehicles exchange their sensor information, such as camera and radar data, which allows other vehicles "see" with the eyes of others and detect otherwise hidden objects);
3. Cooperative Driving phase (vehicles share their trajectories or planned maneuvers data with other traffic participants, allowing them to accurately predict other traffic participant behavior and optimize their own decisions);
4. Synchronized Cooperative Driving phase (high driving automation level (4 and 5) vehicles exchange and synchronize their driving trajectories to achieve optimal driving patterns).

C-ITS deployment phases	**Phase 1:** Awareness driving	**Phase 2:** Sensing driving	**Phase 3:** Cooperative driving	**Phase 4:** Synchronized cooperative driving
C-ITS services	Basic warning services: • Intersection warning • Emergency vehicle warning • Hazard warning • Etc.	Advanced warning services: • Vulnerable road user warning • Overtaking warning • Etc.	• Roadworks assistance • Lane-merging assistance • Platooning • Etc.	• Cooperative merging • Overtaking assistance • Dynamic platooning • Etc.
Data shared among AVs	Status data	Sensor data Status data	Intention data Sensor data Status data	Coordination data Intention data Sensor data Status data
AV driving automation level	Level 1: Driving assistance			Level 5: Full automation

Fig. 2. C-ITS deployment phases (Sjoberg et al. 2017).

Figure 2 shows the C-ITS deployment phases with corresponding services and data, shared among vehicles, which enables the implementation of these services.

To the best of authors' knowledge, there are no international standards defining the architecture of C-ITS. Thus, the international standard ISO 21217 (Intelligent transport systems – communications access for land mobiles (CALM) – architecture) (ISO 21217 2014), which defines the common architectural framework of intelligent transport systems (ITS), could be adapted for C-ITS. ITS is a system-of-systems, which consist of various traffic participants, such as vehicles, roadside infrastructure, portable devices, control centers, etc., connected via various networking and access technologies including the Internet, public and private networks, Bluetooth, Wifi, cellular technologies, etc. Each of these systems contains a functional entity – ITS station. The ISO 21217 defines the architecture of each ITS station.

2.4   CESAM Method for Modelling System-of-Systems

CESAM is a systems architecture and modelling framework, developed by the CESAMES Association to facilitate modelling complex integrated systems, such as systems-of-systems (CESAM 2017). In CESAM, the system is analyzed, and its architecture models are built using three different perspectives: operational, functional, and constructional (see Fig. 3).

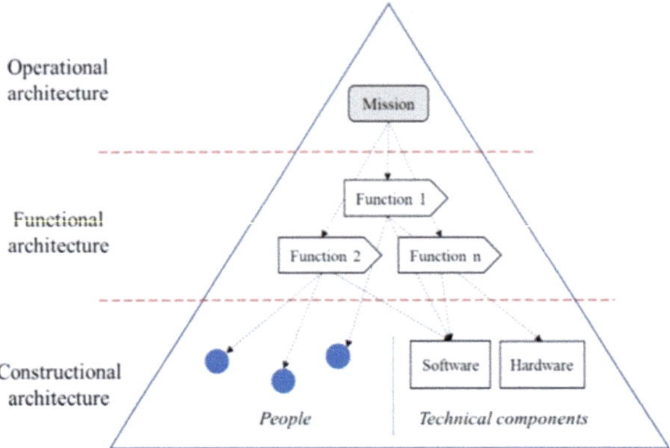

Fig. 3. The CESAM systems architecture pyramid (CESAM 2017).

Operational architecture focuses on understanding the interactions between the system and its stakeholders (environment). As a result of this analysis, the mission of the system is defined, which is an input/output behavior of the environment of the system, involving both the system and its stakeholders. A set of diagrams (need architecture, lifecycle, use case, operational scenario, and operational flow diagrams) is used to describe operational architecture of the system.

Functional architecture is aimed at performing the functional analysis of the system with the goal to describe the functions of a system and relative interactions. The goal of this phase is to understand what system does. The following diagrams are developed during functional analysis phase: functional requirement architecture diagram, functional mode diagram, functional decomposition & interaction diagram (describes the functions and their interactions in a static way), functional scenario diagram, and functional flow diagrams.

Constructional architecture intends to describe the different components of the system and their interactions. The aim of this phase is to understand and specify the detail of the system in terms of the structure. The following are the deliverables of this phase: constructional requirement architecture diagram, constructional mode diagram, constructional composition & interaction diagram (shows system components and their interactions in a static way), constructional scenario diagram, and constructional flow diagram.

As we can see from Fig. 3, the mission of the system, described during operational architecture step, is further decomposed into functions during functional architecture step, and finally into human and technical components during constructional architecture step.

3   Automated Vehicle Safety and Security

The ISO 26262 standard (ISO 26262-3 2011), which defines functional safety for automotive equipment applicable throughout the life-cycle of all automotive Electronic and

Electrical (E/E) safety-related, is currently being used for AV safety analysis. It aims to address possible hazards caused by the malfunctioning behavior E/E systems. The safety process consists of several phases, such as concept, product development, and production, operation, service and decommissioning. Hazard Analysis and Risk Assessment (HARA) is performed during the concept phase, where hazardous events, safety risks and goals are identified. These goals are further refined into the safety requirements, and the safety countermeasures are designed and implemented. Fault tree analysis is commonly used during HARA phase to identify the conditions and events that could lead to high-level hazardous events.

Currently available version of ISO 26262, published in 2011, requires the presence of the human driver to respond to unexpected environments and conditions, and therefore it is not sufficient for highly automated AVs. A new version of ISO 26262, which considers AVs, should be published by the end of 2018.

SAE J3061 is a vehicle cybersecurity standard (SAE J3061 SAE 2016a, b), which was developed using the ISO 26262 standard as a base. Thus, both standards consist of similar phases. Threat Analysis and Risk Assessment (TARA) is performed during the concept phase, where threats, security risks, and security goals are defined. Attack tree analysis is often used for performing TARA. It helps to determine the potential paths that an attacker could take to lead to the top-level threat. ISO and SAE are jointly developing vehicle standard ISO 21434, which will replace SAE J3061 in 2019.

As there are no international standards for defining AV system-of-systems safety and security available yet, we can use 26262 and SAE J3061 for this purpose. In (Sabaliauskaite and Cui 2017) we proposed an approach for AV safety and security analysis at a single-vehicle level, which uses the Six-Step Model in compliance with the international standards SAE J3061 and ISO 26262. In Steps (1) and (2) of the Six-Step Model, AV functional and structural hierarchies are defined and added to the Six-Step Model, along with their relationships. Steps (3) and (4) correspond to AV vulnerability (hazard and threat) analysis. In Step (3), at the end of ISO 26262 HARA phase, the failures are extracted from the fault trees and added to the Six-Step Model. In Step (4), the attacks are extracted from the attack trees at the end of SAE J3061 TARA phase and added to the Six-Step Model. During Steps (5) and (6), safety and security countermeasures, defined during ISO 26262 and SAE J3061 product development phases, are added to the model along with their relationships. See (Sabaliauskaite and Cui 2017) for more details.

4 CESAM&SSM Approach for Modelling Safe and Secure C-ITS

This section proposes an approach for modelling of safe and secure C-ITS, called CESAM&SSM. It integrates the Six-Step Model with the CESAM method. Furthermore, it takes into consideration the C-ITS deployment phases, described in Sect. 2.3.

Why should we integrate the Six-Step Model and CESAM method? Because both of them are *applicable to C-ITS* modelling. Furthermore, they are *similar*, i.e., both are hierarchical complex system modelling methods. Besides, they *overlap*: CESAM functional architecture corresponds to the functional architecture (step (1)) of the Six-Step

Model, while CESAM constructional architecture corresponds to the structural architecture (step (2)) of the Six-Step Model. Finally, they *complement* each other: CESAM is an advanced method for comprehensive complex system modelling, while Six-Step Model – a tool for system safety and security modelling. Thus, their combination can enable comprehensive modelling of safe and secure complex systems-of-systems.

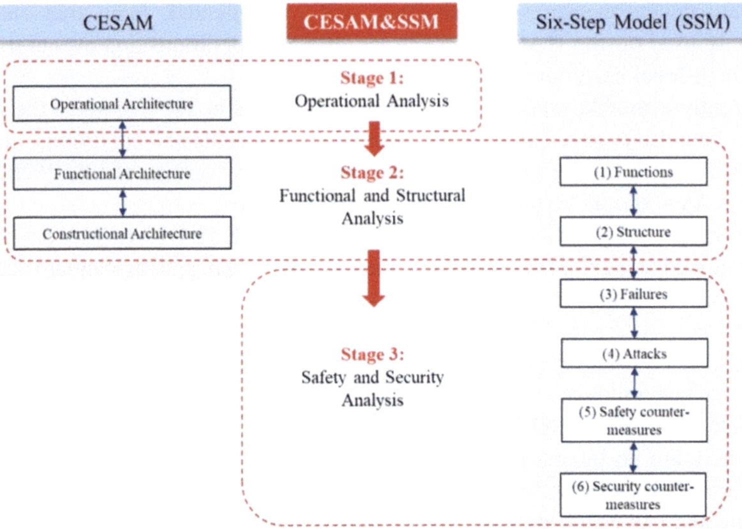

Fig. 4 Safe and secure C-ITS modelling approach CESAM&SSM.

The proposed approach consists of the following three stages (see Fig. 4):

- *Stage 1 – Operational Analysis*. During this stage, the operational architecture of the C-ITS is modelled using the CESAM method, where the interactions between the system and its stakeholders are analyzed and the mission of the system is defined. The following are the C-ITS stakeholders among others: transportation users (drivers, passengers, cyclists, and pedestrians), transportation operators (traffic managers, fleet managers, toll operators, etc.), public safety (incident and emergency management, including police, medical, and fire support), environmental managers (emissions and air quality monitors), Original Equipment vehicle Manufacturers (OEMs), in-vehicle device manufacturers, application developers (in-vehicle, personal device, roadside and control center application developers), communication providers, federal regulators, and policy setting entities (Sun et al. 2016). The mission of the C-ITS depends on the C-ITS deployment phases, and includes the services provided by each phase, as described in Sect. 2.3 (see Fig. 2).
- *Stage 2 – Functional and Structural Analysis*. During this stage, firstly, the functional architecture of the C-ITS is constructed using the CESAM method. Then, the functions and their interactions are extracted from the CESAM functional decomposition & interaction diagram and added to the step (1) of the Six-Step Model. Next, the constructional analysis is performed using the CESAM constructional

architecture step. During this step, the C-ITS station components are analyzed, as defined by the ISO 21217 (2014) (see Sect. 2.3). Finally, the structural elements are extracted from the CESAM constructional composition & interaction diagram and added to the step (2) of the Six-Step Model.

- ***Stage 3 – Safety and Security Analysis.*** During this stage, steps (3)–(6) of the Six-Step Model are performed, where failures, attacks, and safety and security counter-measures are identified using the ISO 26262 and SAE J3061 standards and added to the model along with their relationships with other elements of the model. For more details, see (Sabaliauskaite and Cui 2017).

Models, constructed during stages 1–3, enables the comprehensive analysis of C-ITS safety and security.

5 Conclusions

In this paper, CESAM&SSM – an approach for modelling safe and secure C-ITS is proposed. It integrates the AV safety and security analysis model, Six-Step Model, and the system-of-system architecting method CESAM. A combination of these two methods enables comprehensive analysis of C-ITS from operational, functional, constructional, safety, and security perspectives.

As the CESAM method includes fifteen different diagrams, their construction requires a lot of effort. Thus, we are currently looking for a minimum sufficient set of diagrams to be included in the CESAM&SSM approach.

Furthermore, as the road traffic will never be fully automated, AVs must be able to efficiently and safely coexist with other motorized and non-motorized traffic partici-pants, such as conventional vehicles, pedestrians, and cyclists. Thus, our future research will focus on extending the CESAM&SSM approach for modelling of the mixed traffic systems, which include AVs, infrastructure, and non-automated road users.

References

Behere, S., Törngren, M.: functional reference architecture for autonomous driving. Inf. Softw. Technol. **73**(C), 136–150 (2016). https://doi.org/10.1016/j.infsof.2015.12.008

Brissaud, F., Barros, A., Bérenguer, C., Charpentier, D.: Reliability study of an intelligent transmitter. In: Proceedings of the 15th ISSAT International Conference on Reliability and Quality in Design, San Francisco, USA, 6–8 August 2009 pp. 224–233 (2009)

Car2Car: European vehicle manufacturers work towards bringing Vehicle-to-X Communication onto European roads. Car2Car Communication Consortium Press Release 30 October 2015 (2015). https://www.car-2-car.org/index.php?id=214. Accessed 27 Mar 2018

CESAM: CESAM: CESAMES systems architecting method. A pocket guide. CESAM Community (2017). http://www.cesames.net/wp-content/uploads/2017/05/CESAM-guide.pdf. Accessed 27 Mar 2018

EC: A European strategy on cooperative intelligent transport systems, a milestone towards cooperative, connected and automated mobility. Communication from the Commission of the European Parliament, the Council, the European Economic and Social Committee and the Committee of the Regions. COM/2016/0766 final. The European Commission. Adoption date: 30 November 2016 (2016)

Guardian: Self-driving Uber kills Arizona woman in first fatal crash involving pedestrian. The Guardian (2018). https://www.theguardian.com/technology/2018/mar/19/uber-self-driving-car-kills-woman-arizona-tempe. Accessed 27 Mar 2018

ISO 26262-3 (2011) ISO 26262-3:2011(E), Road vehicles – functional safety – concept phase. The International Organization for Standardization (ISO)

ISO 21217 (2014) ISO 21217:2014(E), Intelligent transport systems – communications access for land mobiles (CALM) – architecture. The International Organization for Standardization (ISO)

Modarres, M., Cheon, S.W.: Function-centered modeling of engineering systems using the goal tree-success tree technique and functional primitives. Reliab. Eng. Syst. Saf. **64**(2), 181–200 (1999)

NHTSA: Automated Driving Systems (ADS): A vision for safety 2.0. The U.S. Department of Transportation and the National Highway Traffic Safety Administration (NHTSA) (2017). https://www.nhtsa.gov/press-releases/us-dot-releases-new-automated-driving-systems-guidance. Accessed 27 Mar 2018

Piètre-Cambacédès, L., Bouissou, M.: Modeling safety and security interdependencies with BDMP (boolean logic driven markov processes). In: Proceedings of the 2010 IEEE International Conference on Systems, Man and Cybernetics, pp. 2852–2861, October 2010

Sabaliauskaite, G., Cui, J.: Integrating autonomous vehicle safety and security. In: Proceedings of the 2nd International Conference on Cyber-Technologies and Cyber-Systems (CYBER 2017), Barcelona, Spain, 12 –16 November 2017 (2017). ISBN 978-1-61208-605-7

Sabaliauskaite, G., Adepu, S., Mathur, A.: A six-step model for safety and security analysis of cyber-physical systems. In: Havarneanu, G., Setola, R., Nassopoulos, H., Wolthusen, S. (eds.) Critical Information Infrastructures Security, CRITIS 2016. Lecture Notes in Computer Science, vol 10242. Springer, Cham (2017)

SAE J3016: SAE J3016, Taxonomy and definitions for terms related to driving automation systems for on-road motor vehicles. SAE International (2016a)

SAE J3061: SAE J3061, Cybersecurity guidebook for cyber-physical vehicle systems. SAE International (2016b)

Sjoberg, K., Andres, P., Buburuzan, T., Brakemeier, A.: Cooperative intelligent transport systems in Europe: current deployment status and outlook. IEEE Veh. Technol. Mag. **12**, 89–97 (2017)

Sun, L., Li, Y., Gao, J.: Architecture and application research of cooperative intelligent transport systems. Procedia Eng. **137**, 747–753 (2016). https://doi.org/10.1016/j.proeng.2016.01.312

Detection of Teamwork Behavior as Meaningful Exploration of Tradespace During Project Design

Puay Siang Tan[✉] and Bryan R. Moser

Massachusetts Institute of Technology, 77 Massachusetts Avenue, Cambridge, MA 02139, USA
puaysiang@alum.mit.edu, bry@mit.edu

Abstract. The increasing complexity of Systems requires Teams of Teams (TofT) from different functional domains to work together. This research aims to better detect and understand the teamwork behaviors and interactions amongst the TofT. Using the *Project Design* approach, 19 groups participated in a model based simulation experiment to reduce the cost and duration of an implementation project for an autonomous vehicle. The performance of the groups was ranked based on generation of non-dominated (cost and schedule) plan alternatives as they explored and simulated the project model. Indicators for coherence in the decision-making process of the groups were explored by the means of visualizing the meaningful exploration of the tradespace via tree diagrams and a "chunking" process. However, sensors to detect meaningful exploration as proposed were not consistently indicative, leading to recommendations for future work on measurement of team exploration and learning during project design.

1 Introduction

Technology advancement has allowed technical systems to be interoperable with other systems to deliver increased functional value. A complex Systems of Systems (SoS) requires a diverse group of domain expertise, leading to the teamwork of multiple functional groups and inter-related agencies in order to deliver the right function and value for user(s). As such, during the development process, tradeoffs by "human(s)-in-the-loop" are critical in addition to classic methods for technical systems optimization. Integrated and higher performance is usually observed in teams with better relationships amongst the various functional groups. In layman terms, the "chemistry" of these groups can led to better project outcomes. Such "chemistry" seems to be episodic in nature and studies in teamwork behaviors explore how to effectively enable the "chemistry" of a Team of Teams (TofT).

This research aims to better detect and understand the teamwork behaviors at the *meso-scale* (inter-functional groups/teams), which is the intermediate level between the *micro-scale* (individuals and small teams) and the *macro-scale* (total portfolios and the firm).

© Springer Nature Switzerland AG 2019
M. A. Cardin et al. (Eds.): CSD&M 2018, AISC 878, pp. 73–87, 2019.
https://doi.org/10.1007/978-3-030-02886-2_7

2   Literature Review

Related research discusses the complexity of technical systems and introduces the concept of sociotechnical systems by looking at roles of mental models, awareness, and attention of the humans interacting with the technical systems. Organizational learning and the role of dialogue for better team performance are discussed. A modelling approach using Project Design is presented to explain how an engineering project can be modelled as a sociotechnical system, thus enabling a model-based design of team behaviors and the measurement of team performance.

2.1   Complex System and Concept of Sociotechnical Systems

The complexity in systems is two-fold – the technical system and social aspect where humans are designing, developing and using the said system. According to Sinha and de Weck, the Structural Complexity, C, of a technical system is measured by the equation: $C = C_1 + C_2 \cdot C_3$, where $C_1 =$ Component Complexity, $C_2 =$ Interface Complexity, and $C_3 =$ Topological Complexity [1]. In this research, the technical complexity of the system is taken as a given and focuses on the various teams working together which forms the "global team" collaborating to realize the complex technical systems i.e. Teams of Team (TofT).

The concept of flexibility in engineering design can be used to deal with uncertainties brought about by topological complexity [2]. Project success requires systems thinking and implementing techniques to explore and retain flexible design. A system viewed as complex is no longer static. Individual decisions may appear safe and rational within the context of the individual work environments and local pressures, but may be unsafe when integrated and operated as a whole. Thus, when a team makes a decision to meet their local conditions, they may be unaware of the potentially dangerous side effects of their behaviors [3]. Complexity can lead to an emergent non-linearity in system responses, often present as project surprises since such responses are not anticipated by experience based know-how nor traditional engineering and management tools.

This research asserts that consideration of team behaviors during complex problem solving can be improved by detecting the events of the problem space (manifested as needs, values, stakeholders), solution space (developed as requirements, function and form of the technical system) and the social space (represented by roles, capability, motivation and power of the actors of the project team). A sociotechnical event (e.g. a decision made by project team during problem solving) will have effects on all 3 dimensions of the problem space, solution space and social space [4, 5].

2.2   Team Performance, Organization Learning and Culture

Organizations assemble teams comprising individuals with different expertise and knowledge to work on complex systems. In organizational theory, firms are conceptualized as systems that "processes" information. In a competitive environment, the organization is faced with how efficiently it can deal with information and decisions in an uncertain environment [6]. While widely accepted within the social sciences that efficiency in organizations is achievable with high performing teams and learning organizations [7], the underlying mechanisms of high performing teams and what metrics measure organization learning remain less clear. For example, Google's Project Aristotle reviewed academic studies on how teams work and yet did not establish strong patterns to define the "ideal" makeup of the team that achieves best effectiveness [8].

Effective organizational learning has been shown as "double loop learning" in which employees are capable of not only detecting errors but also questioning the underlying policies and goals of the organization [7]. Further, organizational learning has been conveyed as knowledge creation [6] in which socialization modes are critical for collective learning, during which members of team shares experiences and perspectives. Secondly, a successive round of meaningful dialogue triggers a mode of externalization. The dialogue may allow team members to articulate their own perspectives and reveal hidden tacit knowledge (i.e. one's understanding and belief of how things should be). Combined with existing data and external knowledge, teams can enter an iterative process of trial and error until they develop a shared concept. The interactions between the tacit knowledge and explicit knowledge tend to become larger in scale and faster in speed as more actors in and around the organization become involved.

The culture of a group can be defined as "a pattern of shared basic assumptions that was learned by a group as it solved its problems of external adaptation and internal integration, that has worked well enough to be considered valid and, therefore, to be taught to new members as the correct way to perceive, think, and feel in relation to those problems" [9]. Culture is divided into three levels: artifacts, values and underlying assumptions. While underlying assumptions might start out historically as values, as they are used and tried over time, they became assumptions and are taken for granted. The danger for the organization is that assumptions might not be unlearnt even though the environment has changed.

2.3 Modeling Approach to Measure Team Performance

The bounds of rationality can be increased by modelling and analysis of engineering systems, extending the cognitive limit of human minds and exploration given finite amount of time in decision making [10].

The Virtual Design Team (VDT) model was used to simulate the behaviors of individuals in information processing, communication, and coordination within a project organization and also predicts several measures of individual and project-level performance [11]. Subsequently, a computational experimental design based on VDT was used to simulate combinations of US vs Japanese organizational style and micro-level behavioral patterns to predict work volume, cost, schedule, and process quality outcomes. The model aims to help managers to find the optimal organizational style to match their project characteristics and teams' micro-behaviors for better team performance [12].

Project Design is an integrative approach through which an engineering project is modeled as a system that captures the interrelatedness of project elements (i.e. product, process and organization) [4, 13]. Since complex projects usually consist of many cross functional teams, these teams bring along their embedded assumptions and practices, thus it is critical to enable these teams to foresee the consequences of their own behaviors and make adjustments accordingly. The project design approach allows the participants to engage in an iterative and social process to evaluate choices and outcomes in a rapid manner, through dialogue, analysis and prototyping, during which awareness can be built. A comparative study has been done to suggest that a team's method for tradespace exploration matters and that influences the shared awareness among team members in early phases of the project life cycle. The model-based and simulation methods are preferred over traditional project planning such critical path and Gantt Charts in order to observe team behaviors during early planning and the tradespace exploration [14].

3   Research Approach

Structured dialogue during which communications are promoted or inhibited are proposed as a means to observe behaviors of a TofT. By exposing these interactions, this research hopes to reveal whether or not these behaviors can lead to different emergent project outcomes (e.g. better team performance, effective team strategy, recognition and response to surprises). This research follows the *Project Design* approach [5] to establish indicative "sensors" to measure teamwork behaviors quantitatively at the mesoscale. In this experiment, the teams are organized in a workshop to participate in a Project Design challenge for the development of an autonomous vehicle, using Team-Port software, a model-based project simulation toolset. The platform allows the teams to make changes to the preset elements of the autonomous vehicle project and perform simulations to get a forecast of the **cost** and **duration** of the project which are defined as **project outcomes**. The participants can <u>make changes to the project model</u>, <u>explore other changes in a new project model</u> and <u>go back to the previous models</u> they have created. This allows them to rapidly explore the tradespace and cover a wide range of possible outcomes within the allotted time for the experiment.

For the experimenter, TeamPort software provided unobtrusive measurements to gain insights on the teams such as their **attention allocation** via the <u>*Fingerprint Reports*</u> (number of clicks on the project elements), **decisions made** via the <u>*Change Log*</u> made in each evolving model instance and related simulation, and the **in-the-moment interactions** via <u>*timestamp data*</u> for correlation of Fingerprints, Change Log and simulation outcomes shown in a tradespace.

Figure 1 showed the matrix of the tools used for data collection. The Design Group's attention and decisions and the act of suspending judgement are considered teamwork behaviors. Presence of strategy during the tradespace exploration, encountering project surprises during the tradespace exploration and the project outcomes (in cost and duration for each project model) is taken to be the emergent effects of teamwork behaviors.

Tools for Data Collection		Demographics	Teamwork Behaviours				Emergent Effects of Teamwork Behavior		
			Attention	Decision	Challenging Assumptions	Suspending Judgement	Strategy	Project Surprises	Project Outcomes
Qualitative	Pre Workshop Survery	✓							
	Post Workshop Survey	✓			✓	✓	✓	✓	
	Scratchpad			✓					✓
Quantitative	Tradespace Report								✓
	Fingerprint Report		✓						
	Change Log			✓					

Fig. 1. Matrix showing qualitative and quantitative tools for data collection.

4   Experiment Implementation

The experiment was conducted with mid-career professionals who participated in a 120 min Project Design Challenge to reduce the cost and schedule of a project for the implementation of an autonomous vehicle. The project was modelled in TeamPort, preset with 3 Work Locations, 13 Teams, 11 Products, 33 Activities, and 7 Phases. Each Design Group began with this same baseline, where the cost and schedule was at **1,037 days** and **$12, 878,438** respectively. Figure 2 shows a view in the TeamPort software

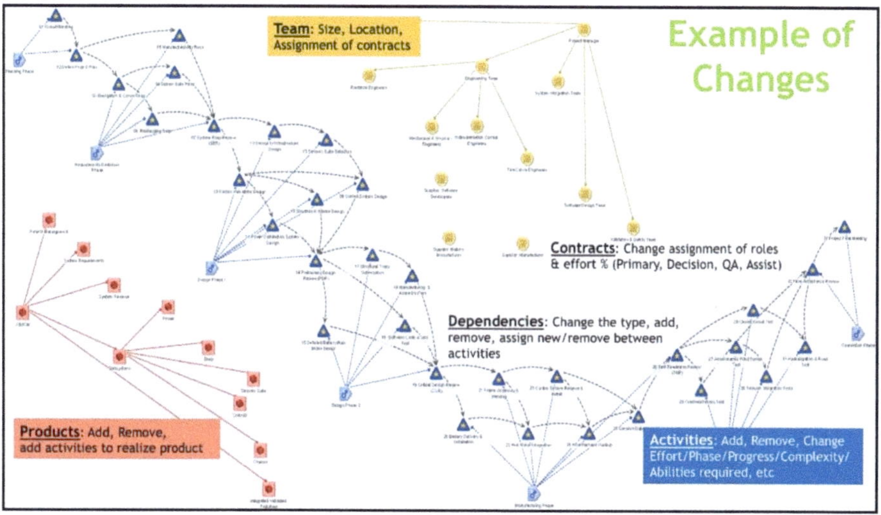

Fig. 2. Architecture view in TeamPort software, overlaid with examples of changes.

and some of the changes that can be made to the project model (i.e. Team, Products, Activities, Dependencies, Contracts).

Data was collected for 19 Design Groups (83 participants) which have 3 to 5 members per group. Table 1 showed the completeness of the data collected.

Table 1. Summary of completeness of data collected

Data collection (of 19 Design groups)	Qualitative data			Quantitative data		
	Pre survey	Post survey	Scratch pads	Trade space	Finger prints	Changes
Total collected	12	14	18	19	11	19
% Completeness	63%	74%	95%	100%	58%	100%

5 Data Analysis

In order to measure team performance, a method is developed to rank the Design Groups' performance. During the Project Design Challenge, 529 project outcomes were simulated by the 19 Design Groups over the 120 min as they explored the tradespace of cost and duration. A Pareto Frontier is first drawn to identify the range of outcomes which are the non-dominated solutions for the 529 data points. A non-dominated solution (i.e. project outcome in this experiment) is not out-performed by any other solution in both cost and duration. The set of outcomes that lie on this first Pareto Frontier are then removed from the tradespace so that a second Pareto Frontier can be identified and the

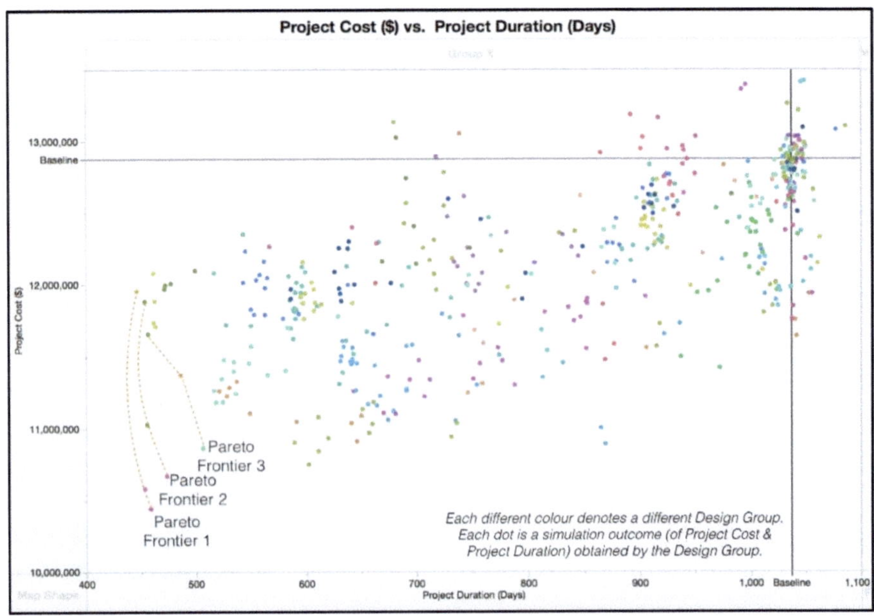

Fig. 3. Illustration of the first three pareto frontiers drawn for tradespace exploration.

process is repeated until all outcomes are sorted into their respective Pareto Frontiers. A total of 66 Pareto Frontiers were derived, Fig. 3 illustrates the first 3 Pareto Frontiers obtained. The project designs for each Design Group are represented with a single color; these points on the tradespace are the simulation outcomes of the Design Groups. The original shared baseline plan can be seen in the upper right.

The ranking of teams using this approach is relative and based on the number of outcomes a Design Group has on the better Pareto Frontiers. A Design Group with outcome in an earlier Pareto Frontier is ranked higher; and for more than one Design Groups with outcomes on the same Pareto Frontier, the Design Group with higher number of outcomes is ranked higher. Once accorded a ranking position, the number of outcomes the Design Group has is no longer used to compare with the new Design Groups appearing on the Pareto Frontier. If there is a tie in a Pareto Frontier, these Design Groups are "on-hold" for the ranking positions until a tie breaker occurs in the subsequent frontiers. If a new Design Group appear in a subsequent frontier which consists of Design Groups which were previously "on-hold" for ranking positions, this new Design Group is ranked lower than those "on-hold" ranking positions.

5.1 Steps in Data Analysis to Explore Indicators to Team Performance

Given the relative ranking of team performance based on the method above, further data analysis is performed to investigate and explore indicators to measure team performance, including *coherence in team decision making* as an indicator of positive team behavior. The components of a proposed coherence analysis are shown in Fig. 4.

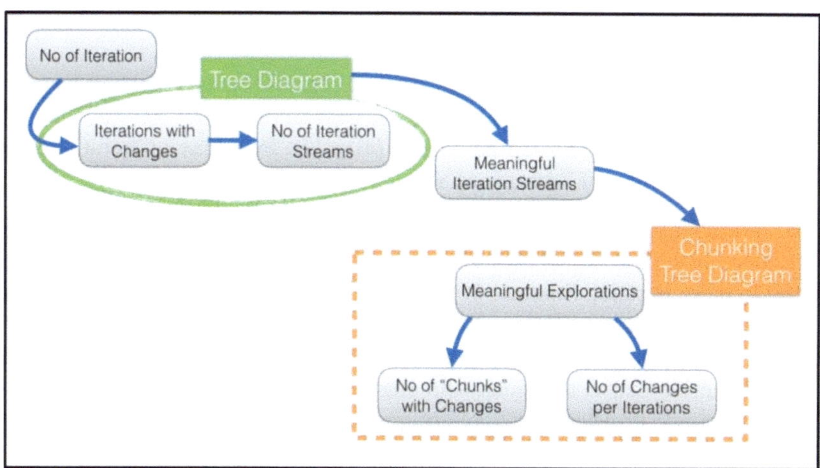

Fig. 4. Data analysis to explore indicators for team performance.

5.2   Visualizing Tradespace Exploration: Tree Diagrams & Chunking Process

The *Number of Iterations* was first used to explore Indicators for Team Performance and the correlation was found to be weak. It was then observed that the iterations performed by the Design Groups followed a "chain-like" pattern as they moved from one Project Model to another. Tree Diagrams were generated to visualize the tradespace exploration of the Design Group by mapping the iterations performed in sequence (henceforth referred to as Iteration Streams) and provided details on the changes type made to the project model (e.g. changes in Team size denoted by Yellow colored circles, changes in Dependencies are denoted by Cyan colored circles).

To consider ways of observing coherence, the differences between exploration by Design Groups with the best and worst outcomes are discussed. See Fig. 5 for the Tree Diagrams contrasting the Top Performing Group and the Lowest Performing Group. It can be observed that the Bottom Performing Group has more "Squares" as compared to the Top Performing Group i.e. the Bottom Performing Group spent more attention coming up with many project models yet did not simulate an outcome (plan forecast with cost and duration).

Secondly, there are many Grey "Circles" and "Squares", meaning that the Bottom Performing Group did not make many changes along those Iteration Streams. In addition, it can also be observed that many of the Iteration Streams had only 1 or 2 steps before the group reverted to the base model (Project Model ID11). What can be inferred is that the group might not have been consistent as they made changes to the Project Models during the tradespace exploration.

The main contrasts of the two Tree Diagrams are:

(1) The Top Performing Group had a high percentage of *Iterations with Changes* over *Total Iterations* as compared to the Bottom Performing Group.
(2) While the Top Performing Group also has Iterations Streams that consist of only 1 to 2 steps, there are also *Iteration Streams with many steps*.
(3) The Top Performing Group has many *branches* from one of Iteration Streams (resulting in a total of 7 Iterations Streams from Model ID69).

The insight suggests that the Iteration Streams in the Top Performing Group show a more coherent exploration of the tradespace especially given the several branches that give rise to more Iteration Streams. While both groups in Fig. 5 have the same number of Iteration Streams, they differ in this meaningful way.

Hence, this research defines *Meaningful Iteration Streams* to be an Iteration Stream that contains more than 5 iterations with changes. Meaningful Iterations which originated from a single Project Model are grouped into a "Meaningful Exploration" of the tradespace. In Fig. 5, the Top Performing Group has 2 Meaningful Exploration, and the Bottom Performing Group has no Meaningful Explorations.

A Chunking Process is then developed to analyze each Meaningful Exploration by segmenting each into "Chunks" which were further classified into blocks with specific number of changes. Figure 6 illustrates the visual output of the "Chunking" Process performed for the Meaningful Exploration-1 of the Top Performing Design Group.

There are 4 "Chunks" with no net changes (Grey), 8 "Chunks" with only 1 change (Purple), and 1 "Chunk" with 4 changes (Blue). That means there are a total of 18 "Chunks" where changes were made, with 14 "Chunks" that had meaningful changes (the remaining 4 "Chunks" had no net changes).

It is also possible to see how many model changes were made within one iteration. This research has defined an iteration with two or less changes per project model as a "Focused"

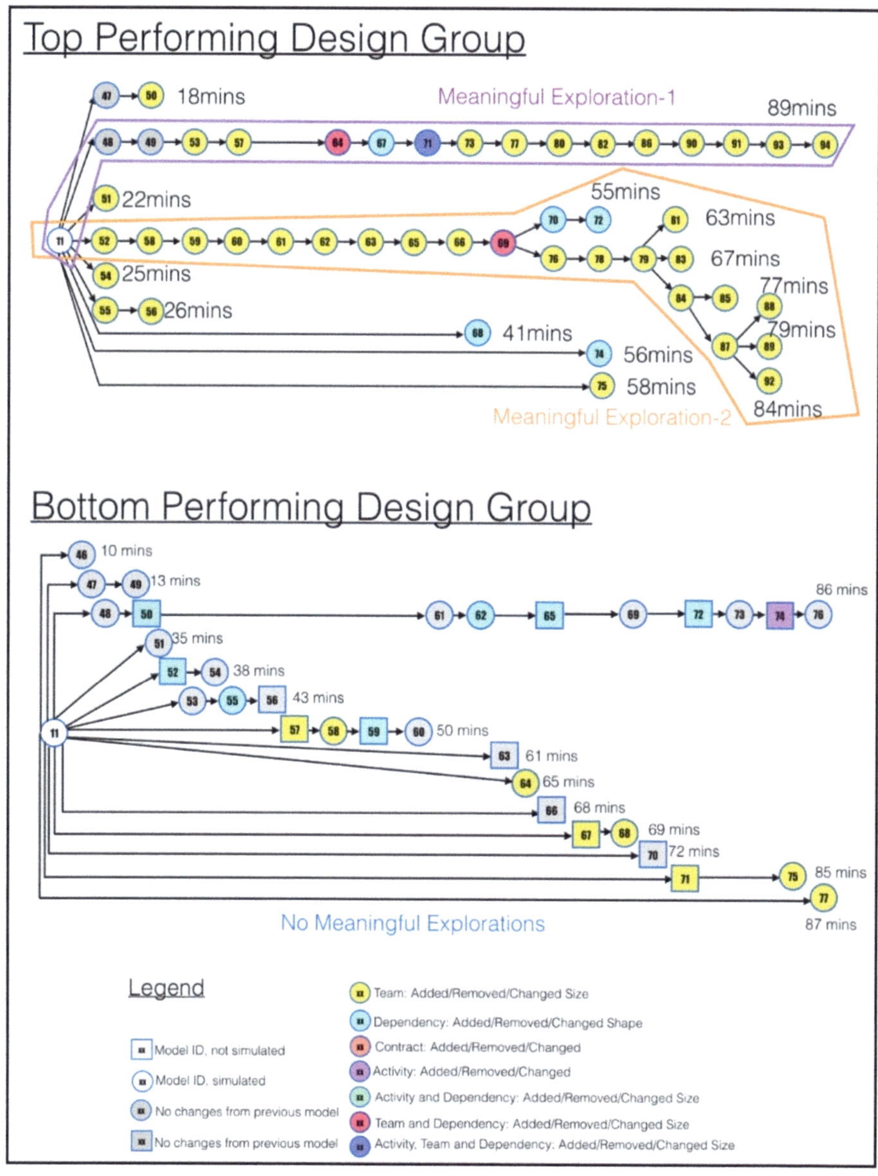

Fig. 5. Tree diagrams of top performing group vs bottom performing group.

82   P. S. Tan and B. R. Moser

Iteration. Figure 6 also showed that Project Model ID 64 and 71 are not considered "Focused" Iterations as there were 3 changes and 5 changes respectively in each iteration.

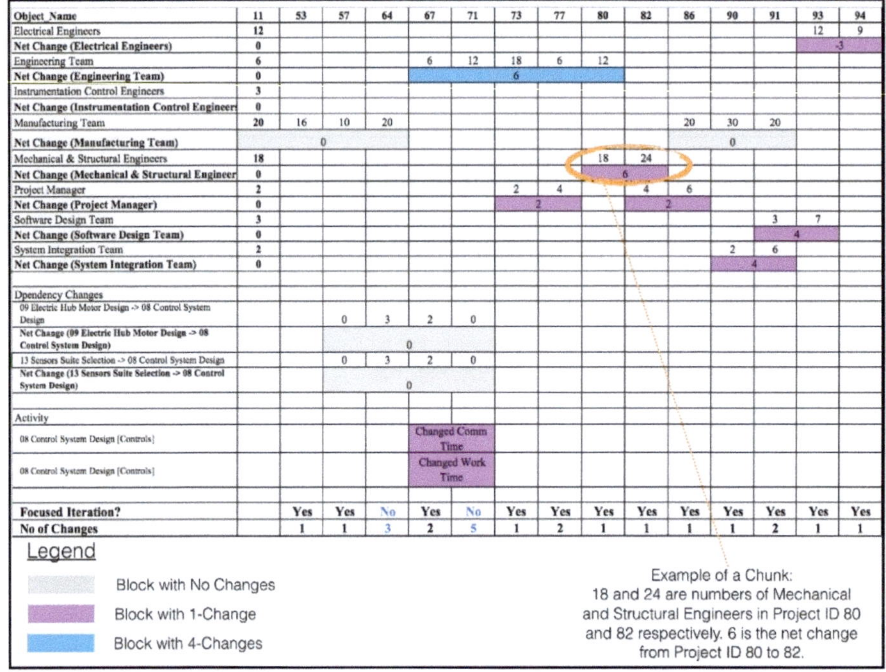

Object_Name	11	53	57	64	67	71	73	77	80	82	86	90	91	93	94
Electrical Engineers	12													12	9
Net Change (Electrical Engineers)	0														-3
Engineering Team	6				6	12	18	6	12						
Net Change (Engineering Team)	0						6								
Instrumentation Control Engineers	3														
Net Change (Instrumentation Control Engineers)	0														
Manufacturing Team	20	16	10	20							20	30	20		
Net Change (Manufacturing Team)	0		0									0			
Mechanical & Structural Engineers	18								18	24					
Net Change (Mechanical & Structural Engineers)	0									6					
Project Manager	2						2	4		4	6				
Net Change (Project Manager)	0							2			2				
Software Design Team	3												3	7	
Net Change (Software Design Team)	0													4	
System Integration Team	2											2	6		
Net Change (System Integration Team)	0												4		
Dependency Changes															
09 Electric Hub Motor Design -> 08 Control System Design			0	3	2	0									
Net Change (09 Electric Hub Motor Design -> 08 Control System Design)					0										
13 Sensors Suite Selection -> 08 Control System Design			0	3	2	0									
Net Change (13 Sensors Suite Selection -> 08 Control System Design)					0										
Activity															
08 Control System Design [Controls]					Changed Comm Time										
08 Control System Design [Controls]					Changed Work Time										
Focused Iteration?		Yes	Yes	No	Yes	No	Yes	Yes	Yes	Yes	Yes	Yes	Yes	Yes	Yes
No of Changes		1	1	3	2	5	1	2	1	1	1	1	2	1	1

Fig. 6. Output of chunking process for meaningful exploration-1 of top performing group.

5.3   Investigating Indicators for Team Performance

The various indicators for team performance are investigated with respect to the ranking positions of the Design Groups (see Figs. 7, 8).

In Fig. 7, it can be observed that the Number of Iterations with Changes (Orange) has a similar trend as the Total Number of Iterations (Blue) except for the Worst Performing Group (Ranked #19). Further comparison is made with Number of Iteration Streams (Green) and Number of Meaningful Iterations Streams (Yellow). It was observed that the Worst Performing Group has high number of Iteration Streams and yet zero Meaningful Iteration Streams, thus suggesting the Number of Meaningful Iteration Streams may be a better indicator for team performance than Number of Iteration Streams.

Following the above insight, Fig. 8 presents the plot of Number of Meaningful Iteration Streams against the ranking positions. A downward trend is observed especially for the Bottom 10 Performing Groups. This trend may suggest that the Number of Meaningful Iteration Stream is a better indicator for Team Performance as compared to Number of Iteration Streams, Number of Iterations with Changes or Total Number of Iterations.

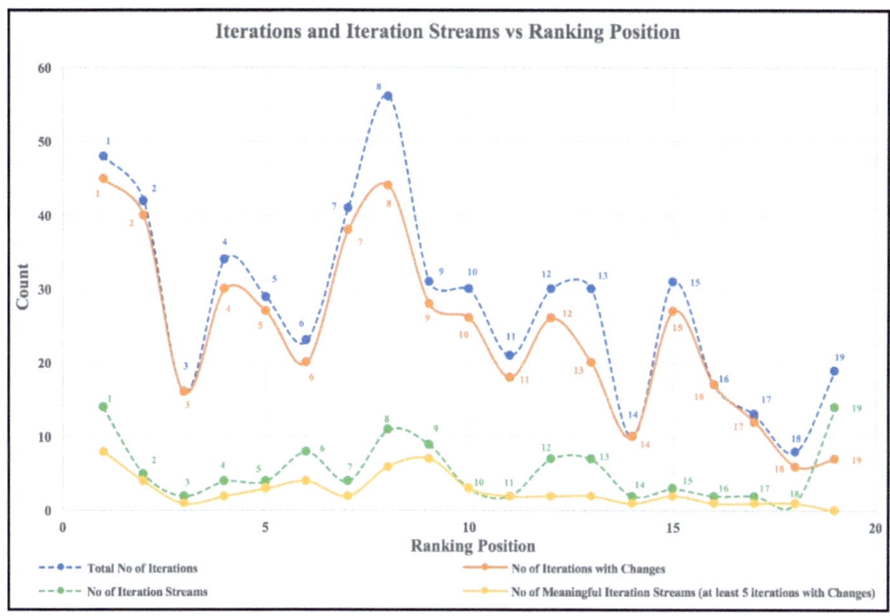

Fig. 7. Comparing iterations, iterations with changes, iteration streams and meaningful iterations streams with ranking positions.

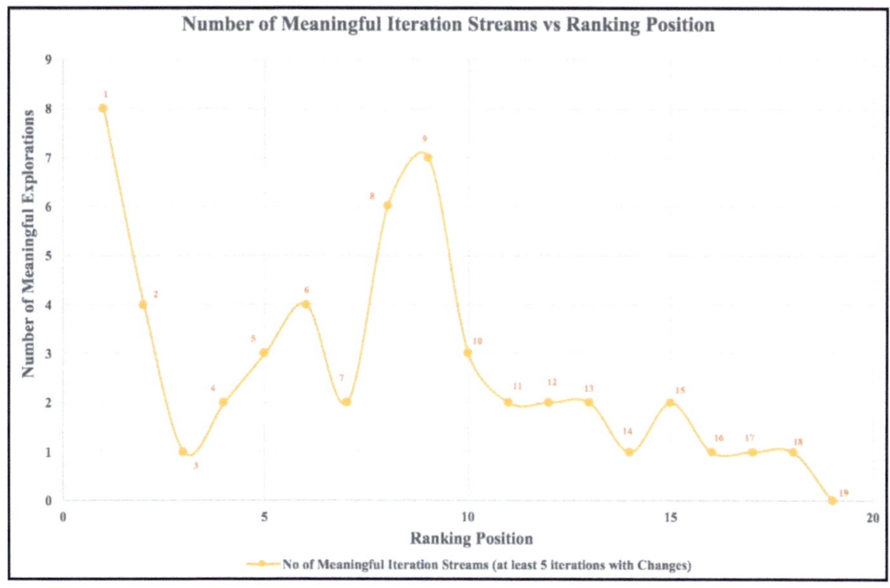

Fig. 8. Meaningful iteration streams vs ranking positions.

5.4 Conclusions from this Initial Experiment

In this experiment, the Number of Meaningful Iterations Streams may be a better indicator of Team Performance in which Meaningful Explorations of the tradespace can be observed and analyzed. Furthermore, we propose measuring how focused a Design Group is during the decision making process, which we refer to as *Indicators of Coherence*, such coherence may signal a better chance of meaningful changes made during tradespace exploration.

6 Discussions and Insights

6.1 "Meaningful Explorations" to Detect Presence of Strategy Quantitatively

Meaningful Explorations were defined to chunk the Tree Diagrams to develop indicators for coherence. Coherence of the tradespace exploration is used as a proxy in this research to detect presence of strategy, and thus indicative of Team Performance.

Several ratios of the indicators of performance above were proposed as *indicators of coherence*. (For example, the ratio of *Focus Iterations* to *Total Iterations with Changes*). These ratios used the quantitative data from the Chunking Process and were bolstered by qualitative data collected through self-reporting surveys by the Design Groups (e.g. "Did your team follow a strategy?"). However, these ratios as proposed were not indicative for this experiment, suggesting no correlation between the initially proposed coherence ratios and the team performance order. Several different ratios and combination functions of positive and negative indicators will be explored in future work.

6.2 Framework for Real Time Feedback on Team Performance

One of the challenges that general managers face is how to evaluate different teams in a fair and objective manner. Team Performance evaluation can be highly subjective given that each team works on projects of different complexity and deliver project outcomes measured in different value.

The research has established the following framework:

1. Systematically rank team performance upon completion of a Project Design Challenge
2. Use the Tree Diagram to visualize the tradespace exploration
3. Explore indicators of coherence during tradespace exploration based on the Chunking Process.

With the framework, similar experiments can be brought to real-world industry to provide real-time feedback on team performance in organizations.

6.3   Detecting Presence of Project Surprises

Project surprises are outcomes of the system performance unexpected by the teams. For example, a team had anticipated a certain outcome (or even a range of outcomes) when they implemented parameter changes, given the assumptions and experiences they have with the system. However, the system responded in a manner outside expectations. For example:

- Negative Surprises – Adding more team members leads to increase (instead of decrease) in project duration and costs.
- Positive Surprises – Removal of team waiting for work leads a very significant reduction of costs and schedule.

Mistakes in implementation (i.e. where assumptions and understanding of the system are valid yet emergent systemic effects lead to a different system performance outcome, not overturning the existing assumptions of the team) are excluded from the considerations of project surprises. In this experiment, mistakes in implementation refers to the iterations of a Design Group yet not a meaningful exploration, which is likely to happen during early stage of the experiment.

In this experiment, qualitative data are collected to explore if a Design Group has encountered project surprises via the Post Workshop Survey and a Scratchpad. The responses from the Post-Workshop surveys indicate if individuals within the same Design Group had different beliefs and when the project surprises were encountered. As for Scratchpads, some Design Groups did not complete their entries required. Thus, the responses collected did not have sufficient sample size and were not used for analysis.

Nevertheless, the following response examples suggest that indeed the Design Groups did encounter project surprise lends supportability that it is possible to define indicators for surprises in future experiments using the Project Design approach, for example:

- Moving teams to India, adding more members would help most times, but there was a definite "art" to it; sometimes things would go the wrong way
- Increasing PM capacity led to a higher performance unexpected by most of the team members. Allowing a test through simulation of the Project Design was key to fight counterintuitive ideas.

7   Future Work

7.1   Instrumented Teamwork Behavior Measurement

In this research, qualitative data was collected to provide quasi-information on teamwork behavior as tools for instrumented measurements of quantitative data were not yet available. Use of qualitative data entails the risk of data being subjected to individual opinions, which include biases and assumptions of the member of the Design Group and potentially not a true reflection of the behaviors and responses of the Design Group (representing the TofT). In addition, the data collected might not be complete for a thorough and fair analysis to be performed as individuals might choose not or fail to respond during the experiment.

Thus, instrumented measurements are proposed to be embedded in the software in which the Design Groups can immediately record responses such as their expected results prior to performing the simulation and indications if they had encountered project surprises when the simulation results turned out to be not what they had expected.

7.2 Measuring Team Learning Effects

With a definitive methodology to rank Team Performance, the next important question in advancing this research delves into: *"What are the Team Learning effects as an emergent property of the Project Design Challenge?"*

A possible Design of Experiment could include a 2-step data collection approach, in which a Project Design Experiment similar to the one conducted in this research is first conducted and subsequently repeated with the same participants (making up the Design Groups) 2 to 3 months later.

It would be extremely valuable to evaluate the Design Group's performance after Step 2, especially if the members of the Design Groups have been working together over the 2 to 3 month period, in which they might have developed a shared mental model. The data collected is likely to advance the knowledge and insights to Team Performance and Learning Effects in Teamwork.

8 Conclusions

In this research, an experiment was conducted which allows for real time feedback to Design Groups on the impact and emergent effects of their actions in making changes to a project model. The research has focused on evaluating team performance. The researchers attempted to detect the presence of team strategy and subsequently explored indicators of coherence to observe patterns of meaningful tradespace exploration. Research work remains on the detection of team learning and its effects. The current progress of this research indicates that it is possible to conduct instrumented teamwork experiments with a Project Design approach to develop indicators and sensors for team learning during complex engineering planning.

References

1. Sinha, K., de Weck, O.: Structural complexity metric for engineered complex systems and its application, pp. 181–192 (2012)
2. Neufville, R.D., Scholtes, S.: Flexibility in Engineering Design. MIT Press, Cambridge (2011)
3. Leveson, N.G.: Engineering a Safer World: Systems Thinking Applied to Safety. MIT Press, Cambridge (2012)
4. Moser, B., Mori, K., Suzuki, H., Kimura, F.: Global product development based on activity models with coordination distance features. In: Proceedings of the 29th International Seminar on Manufacturing Systems, pp. 161–166 (1997)

5. Moser, B.R., Wood, R.T.: Complex engineering programs as sociotechnical systems. In: Concurrent Engineering in the 21st Century, p. 51, January 2015
6. Nonaka, I.: A dynamic theory of organizational knowledge creation. Organ. Sci. **5**(1), 14 (1994)
7. Argyris, C.: Double loop learning in organizations. Harv. Bus. Rev. **55**(5), 115–125 (1977)
8. Duhigg, C.: What Google Learned From Its Quest to Build the Perfect Team. The New York Times, 25 February 2016
9. Schein, E.H.: On dialogue, culture, and organizational learning. Organ. Dyn. **22**(2), 40–51 (1993)
10. Simon, H.A.: Models of Bounded Rationality: Empirically Grounded Economic Reason. MIT Press, Cambridge (1997)
11. Levitt, R.E., Thomsen, J., Christiansen, T.R., Kunz, J.C., Jin, Y., Nass, C.: Simulating project work processes and organizations: toward a micro-contingency theory of organizational design. Manag. Sci. **45**(11), 1479 (1999)
12. Horii, T., Jin, Y., Levitt, R.E.: Modeling and analyzing cultural influences on project team performance. Comput. Math. Organ. Theory **10**(4), 305–321 (2005)
13. Moser, B.R., Wood, R.T.: Design of complex programs as sociotechnical systems. In: Concurrent Engineering in the 21st Century, p. 197, January 2015
14. Iluz, M., Moser, B., Shtub, A.: Shared awareness among project team members through role-based simulation during planning – a comparative study. Procedia Comput. Sci. **44**(Supplement C), 295–304 (2015)

Leveraging Data Analytics in Systems Engineering – Towards a Quantum Leap in Railway Reliability

Thaddeus Tsang[✉], Joyce Hong, Mun Yih Wong, and Kum Fatt Ho

Land Transport Authority, 1 Hampshire Road, Block 10, Level 3, Singapore 219428, Singapore
`{thaddeus_tsang,hong_pek_foong,wong_mun_yih,`
`ho_kum_fatt}@lta.gov.sg`

Abstract. Today's world sees data analytics more prevalent than ever before, mining and interpreting data through various trends. In Singapore, the Land Transport Authority's (LTA) new role as asset owner for its railway operating assets has also given impetus towards establishing a sustainable digital ecosystem. Timely decisions are enabled based on accurate understanding of the asset condition from the network of data and analytic processes, to maintain safety and reliability of the Rapid Transit Systems (RTS) throughout its service life. In addition to expounding on the data analytics process, this paper also explores the potential benefits from the data discovery and the challenges in doing so. The crux is to provide a more robust, reliable and resilient public transport system through data analytics, as we strive towards a quantum leap in railway reliability for our commuters.

1 Introduction

Systems Engineering is an inter-disciplinary field of Engineering, encompassing the design and management of complex systems throughout their life cycles. By providing a holistic systems view, Systems Engineering helps mould all the technical inter-dependencies through iterative and concurrent engineering process to achieve an integrated and optimised system solution as a whole. The V-model is widely used to illustrate a Systems Engineering life cycle from cradle to grave. To tackle the complexity of a Rapid Transit System (RTS) development, the V-model is tailored, as shown in Fig. 1, to focus on the key areas in a RTS life cycle from requirements elicitation, decomposition and definition, integration and verification to integrated system acceptance, operations and maintenance, upgrade and decommissioning.

Coupled with a comprehensive systems engineering process is the need for accurate, timely and accessible information about the system and its components throughout the entire life cycle. With over 60,000 operating asset items in a typical RTS, data analytics, supported by prognostic algorithms on big data collected from the asset operations and maintenance would undoubtedly be increasingly pertinent for a deeper understanding of the asset behavior, interactions and performance. The data mined and analysed over a multi-dimensional spectrum could then be used to support decisions for "faster, cheaper and better" solutions fit for effective asset management strategy [2].

M. A. Cardin et al. (Eds.): CSD&M 2018, AISC 878, pp. 88–97, 2019.
https://doi.org/10.1007/978-3-030-02886-2_8

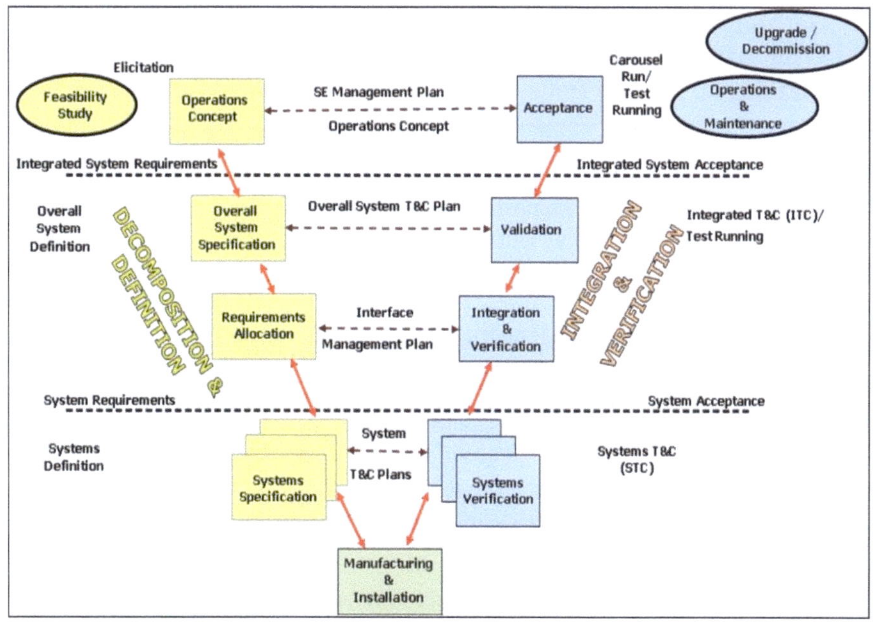

Fig. 1. V-model of a systems engineering life cycle [13]

2   Objective

Under the New Rail Financing Framework (NRFF), the LTA would bear the responsibilities of an asset owner for all rail infrastructure such as stations, depots, and operating assets, including making strategic decisions to build, replace or upgrade the rail assets. Benchmarking against the best-in-class, the Singapore RTS is targeted to achieve a reliability of 1 million Mean Kilometres Between Failure (MKBF)[1] [11]. This is a huge leap in improvement required considering the reliability in 2015 was hovering just over 130,000 MKBF.

Putting LTA's role into perspective, it would comprise first, identification of areas for concern, such as assets where operability has service impact, so that tactical decisions could be made in a timely manner.

The objective of this paper is to illustrate the exploratory work using data analytics on related faults and delays. In particular, Tableau software is used to demonstrate different viewpoints which can be derived from a data set. And this would support comprehension of the operating assets under study.

---

[1] MKBF is the metric used internationally to measure train reliability. 1 million MKBF refers to an average of 1 million train kilometres between delays for >5 min failures.

3   Methodology

3.1   Cleaning up the Raw Data

Raw data collected from operation and maintenance are often disorganised and/or incomplete, especially when there is man-in–the-loop e.g. maintainer's inputs required for on-site observation and diagnosis where advance diagnostic tools for automatic feedback are unavailable. Furthermore, manually sorting of the data using spreadsheets are also prone to human error, potentially masking the true representation of system behavior.

The programming language, R script is used to define a suitable data cleaning algorithm. Although the algorithm may need to be tweaked to interpret data in different formats and structure across various systems in RTS, it automates the process to synthesize and extract meaningful outputs, hence increasing the integrity and trustworthiness of the results. As shown in Fig. 2, new data received can also be effortlessly appended so long as the form is consistent with the template used in the algorithm. Analysis and latest trends can thus be presented promptly for necessary actions.

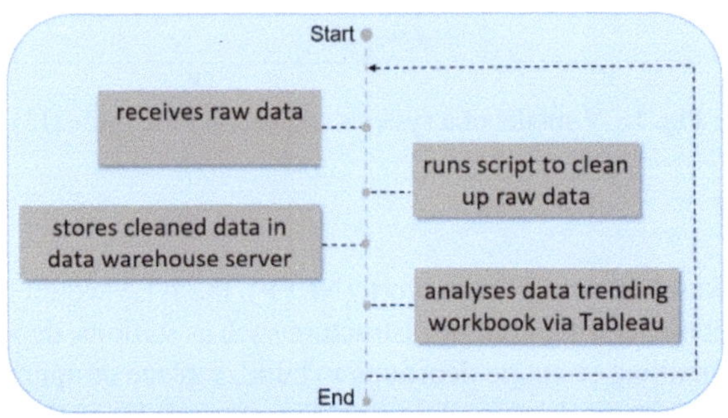

Fig. 2. The data clean up cycle

3.1.1   Analytics Charts for Analyses

Customised trends, such as an overview of systems performance as shown in Fig. 3, can be automatically generated using data visualization tools[2]. Further analysis can be carried out as one manoeuvres with the various dimensions in the dataset to appreciate observations from different viewpoints, using the same set of data. For instance, other key operation parameters such as Location (Trackside/Station/Trainborne), Failure Count, Systems, Sub-systems and Peak/Off-peak Hours may be used to produce additional trending charts. A more holistic health report can then be made on the RTS line under study.

---

[2] Data visualisation tools are digital tools that allow the user to analyse data through the use of smart and even interactive visuals, not limited to charts and graphs. Tableau is an example of a visual analytics software.

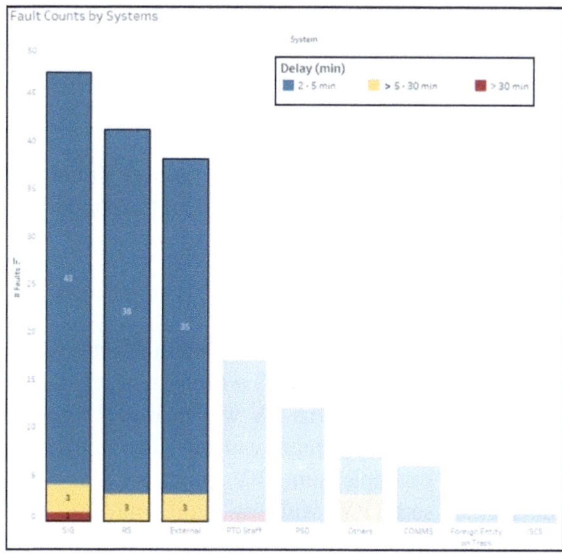

Fig. 3. Failure count by systems

3.1.2 Failure Count by Sub-systems

Data can also be analysed at a higher level of granularity to derive any signs of vulnerability prior to actual failure. For instance, similar failures in the 2 to 5 min delay category may reveal emerging symptoms of a more severe consequence causing longer delay.

Figure 4 demonstrates the potential viewpoint which shows that the failures of some modules, resulting in more than 5 min delay, could indeed be associated with earlier

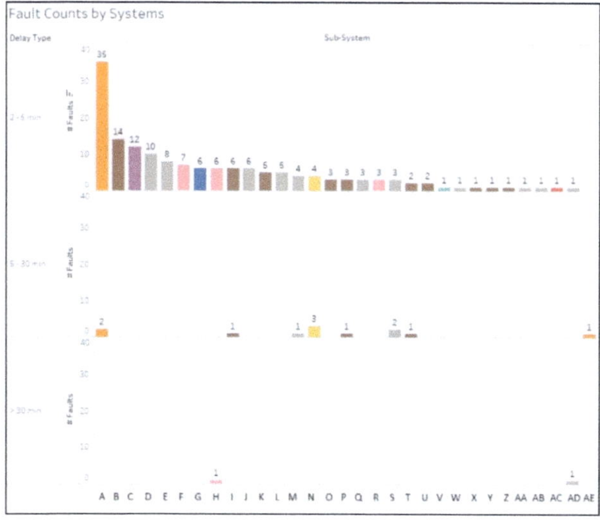

Fig. 4. Failure count by sub-systems

incidents that caused only 2 to 5 min delays. Still, it was interesting to note that there were 2 counts of incidents, resulting in 30 min delays, which could not be traced to any failures in the other categories prior to their occurrence. Nonetheless, such failures should still be considered for update into the reliability critical asset list for close monitoring.

Furthermore, based on the fault classifications, failures within each category can be associated with lower level modules within the system to support more precise, corrective actions.

3.1.3   Failure Count by Passenger Vehicle (PV)

Analysis may also be carried out on a particular system of interest. Figure 5 shows that there was no specific Passenger Vehicle (PV) with particularly high failure frequency. The maximum fault count for a particular PV was merely 9 counts. Neither is there a clear and positive relationship between faults causing 2 to 5 min delay and those causing more than 5 min delay. For example, some PVs with low 2 to 5 min delay count had 30 min delay counts. Also, some PVs with low fault count had contributed to longer service delays upon failure. Hence, looking at the PV total fault count may not be indicative of the focal point. More data could be included for analysis to substantiate more concrete results.

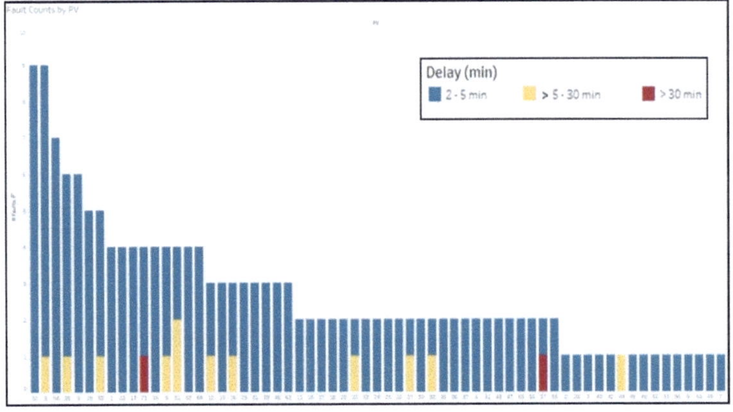

Fig. 5.   Failure count by PV

3.1.4   PV – Location Failure Distribution (PV)

To validate the hypothesis that system behaviour could be affected by the interfacing factors, PV fault count was also plotted against location (train stations) as shown in Fig. 6. The failure distribution appeared to be relatively random, with specific PVs causing failures at the same station, at most, twice within the observation period. Two propositions could be made based on the observed results i.e. failures may not necessarily be due to a particular train at a specific station or location; the emergent properties at the interface has not been revealed due to the presence of other control measures. More data could be collected to support further examination and appropriate proposition.

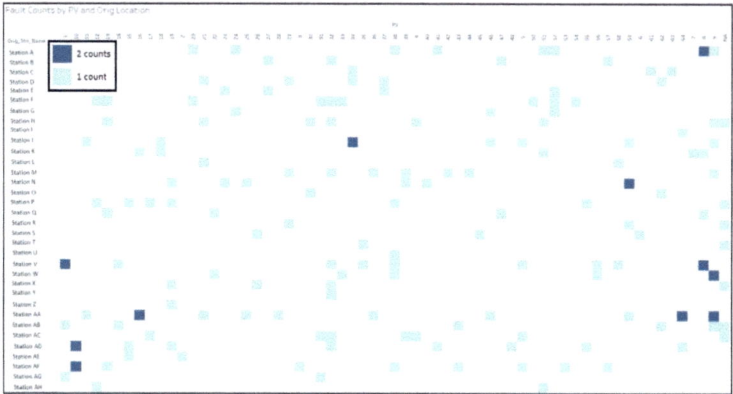

Fig. 6. PV - location failure distribution

3.1.5   System Failure – Location Distribution
The failures at each location were also analysed and plotted against system in Fig. 7 and results showed that:

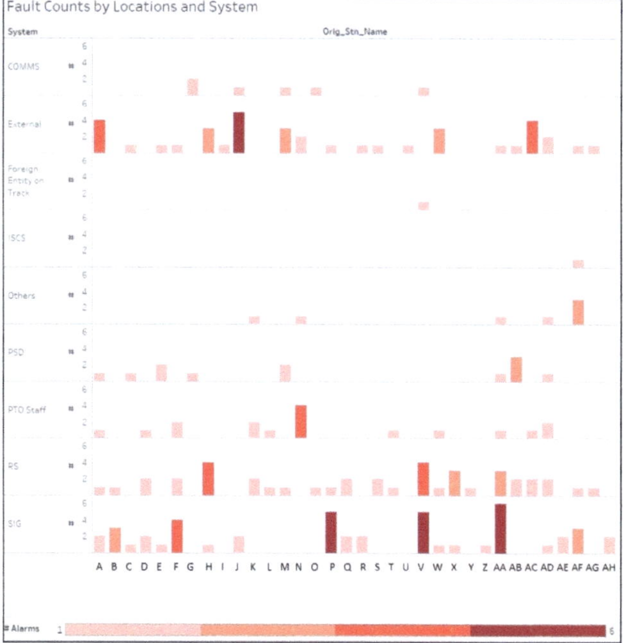

Fig. 7. System failure - location distribution

94   T. Tsang et al.

a. Signalling failures were more prevalent at stations denoted by F, P, V and AA (with at least 4 counts).
b. Rolling Stock failures were, on the other hand, more prone to happen at stations denoted by H, V, X and AA stations with (3 counts or more).
c. PSD failures were highest at station denoted by AB with 3 counts over the observed period).

3.1.6 Failure Count – Location (In Sequence) Distribution

Figure 8 illustrates the failure count based on connectivity between stations in the network. The outcome showed a slightly higher concentration of failures at interchange stations or stations with higher passenger volume. Stations with the most number of failures ranged from 8 to 13 failure counts. Hence, one could infer that failures could be due to heavier loading on the system i.e. loading is a critical factor to asset health.

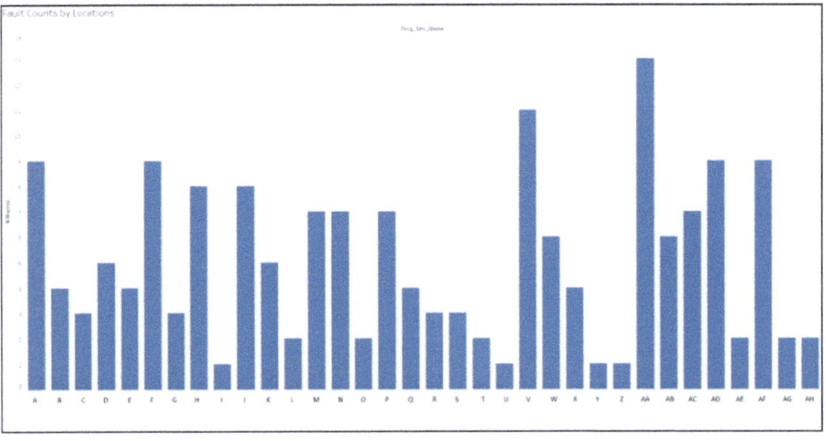

Fig. 8. Failure count - location (In Sequence) distribution

3.1.7 Peak and off-Peak Hours Failure Distribution

Figure 9 shows the comparison between failures during peak and off-peak hour operations. The general correlation seen in the chart was that the more hours covered, the more failure counts there were. Failure counts did not necessarily increase significantly during peak hours. More data could be included to substantiate more concrete findings.

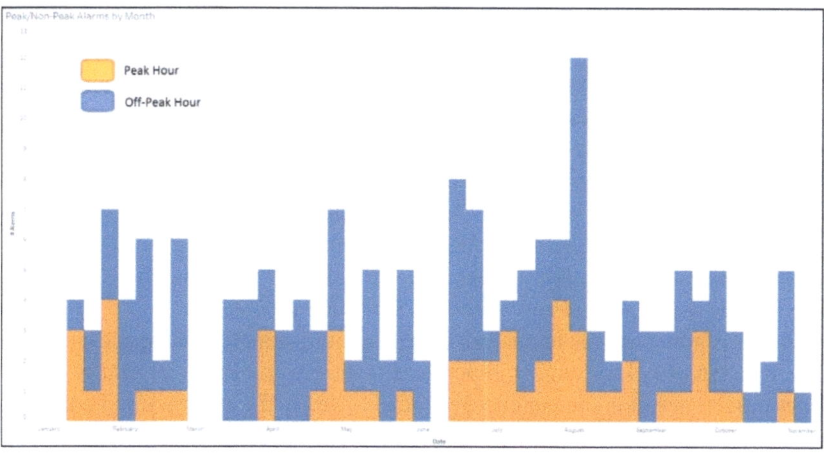

Fig. 9. Failure count - peak/off-peak hour distribution

4   Challenges

The data and various viewpoints can provide insights to system behavior hence facilitate better operation and maintenance strategies. However, data analytics is by no means the silver bullet. Whilst it may be used to interpret voluminous data amid complexity, dataset integrity is the most important piece of the jigsaw. If the data is inaccurate or incomplete, no amount of data analytics applied will provide meaningful results. Where data collection requires human intervention, data integrity would inevitably be at a higher risk. Training and standardisation may reduce the risk. Frequent audits may also help to ensure correctness and completeness in the data collected.

In addition, one will ultimately need to possess the technical know-how on the usage of data analytics in order for it to be useful. The use of data analytics is akin to using any other complex software. For example, Tableau may be a powerful software in visual analytics and can create meaningful visuals beyond the capabilities of typical charts on Excel but it will be completely useless if one knew little of how to exploit it. At the same time, when data gets larger and larger, one would also require the technical know-how to fully grasp the meaning of the data through effective visualisation tools and to discern between possible noise and the signal in the noise [12]. Hence, appropriate training will have to be stepped up to equip one with the relevant knowledge and core skillsets for effective output.

Furthermore, it is also paramount for one to have sufficient domain knowledge to appropriately complement the analysis trends and findings through the elicitation of data mining. This is because not every correlation is meaningful; some could even be mere coincidences and have no direct causal relationship between them. In the world of engineering and design, only one with sufficient technical expertise will be able to discern which correlations do actually have a causal relationship or meaning to the system under study.

5   Conclusion

Data intelligence gathered from increasing density in the network of data sources and growing capabilities on data analytics would inevitably enhance the systems engineering process and asset life cycle management. With the trend analyses, asset owners as well as operator and maintainers could be better informed of specific vulnerabilities in the systems, sub-systems or modules. Acute strategies to operate, maintain, modify, upgrade, redesign or to even dispose/decommission an asset could thus be adopted with more conviction. Furthermore, the insights gained from data analytics can be used to derive lessons learnt and set new baselines for future projects.

While it is now possible to analyse data on a multi-dimensional spectrum like never before, data integrity is of utmost importance to ensure credibility of results and recommendations. Hence, it is fundamentally crucial to verify the accuracy and completeness of any data collected prior to using it. With accurate data, one can now better identify critical systems and sub-systems that are prone to failure, as well as find new leads. With suitable prognostic algorithms, the data may even be used to predict which items could potentially fail prior to occurrence. This would subsequently allow the undertaking of reliability improvement activities like corrective and preventive maintenance much earlier and more efficiently.

Effective Systems Engineering requires the accurate understanding of system behavior and its interactions with other systems within the environment. Analysis of field data under actual operating conditions coupled with timely feedback into the design and development will certainly ease in performance assurance and consequently, future-proofing the system through its useful life. Although confronted with challenges, one can still be confident that the benefits to leveraging data analytics in Systems Engineering for a quantum leap in railway reliability is extensive. And this, would be core to maintaining world class standards in railway.

Acknowledgements. The author would like to sincerely express his gratitude towards the Land Transport Authority (LTA), Singapore in particular, Ms Hong Pek Foong Joyce, Mr Wong Mun Yih and Mr Ho Kum Fatt for their guidance and mentorship in the production of this paper. Their knowledge, experience and discussions were instrumental in exploring the use of data analytics in systems engineering to identify problematic areas/areas worth investigating further in Singapore's RTS as the Land Transport Authority (Singapore) work towards improving railway reliability.

References

1. 4. Systems Engineering. Defense Acquisition Guidebook. Defence Acquisition University, (2013)
2. Forsberg, K., Mooz, H.: System engineering for faster, cheaper, better. Center of Systems Management (1998)
3. ISACA. Data Analytics – A Practical Approach, An ISACA White Paper (2011)
4. The Land Transport Authority (LTA) & SMRT. Circle Line Signalling Problems Caused By Intermittent Failure of Signalling Hardware on Train, A Joint Press Release (2016)

5. Tegos, M.: Data Scientists caught Singapore's 'rogue' train. Here's what else they can do, TECHINASIA (2016)
6. Fai, L.K.: Singapore to increasingly use technology, data analytics to combat transnational crime: DPM Teo, Channel NewsAsia (2017)
7. Tang, J.: Tableau to train Singapore government officers in data science, TechTrade Asia (2017)
8. Tham, I.: 10,000 public servants to receive data science training under GovTech-NUS tie-up, The Straits Times (2017)
9. SMRT Trains' and SMRT Light Rail's Transition to the New Rail Financing Framework: Top 7 Things You Need to Know, Factsheet, Land Transport Authority, (2016)
10. SMRT Trains and SMRT Light Rail To Transit To New Rail Financing Framework, News Releases, Land Transport Authority (2016)
11. Tan, C., Boon, K.: Wan sets new railway network reliability target as MRT becomes three times as dependable as in 2015, The Straits Times (2017)
12. Minelli, M., Chambers, M., et al.: Big Data, Big Analytics: Emerging Business Intelligence and Analytic Trends For Today's Businesses (2013)
13. Joyce, H.P.F., Hin, O.S., et al.: A Systems Assurance Perspective Towards Generic Systems Engineering, INCOSE (2011)

Author Index

© Springer Nature Switzerland AG 2019
M. A. Cardin et al. (Eds.): CSD&M 2018, AISC 878, p. 99, 2019.
https://doi.org/10.1007/978-3-030-02886-2

MIX
Papier aus verantwortungsvollen Quellen
Paper from responsible sources
FSC® C105338

If you have any concerns about our products,
you can contact us on
ProductSafety@springernature.com

In case Publisher is established outside the EU,
the EU authorized representative is:
**Springer Nature Customer Service Center GmbH
Europaplatz 3, 69115 Heidelberg, Germany**

Printed by Libri Plureos GmbH
in Hamburg, Germany